AUTODESK® INVENTOR® 2013 AND
AUTODESK® INVENTOR LT™ 2013

ESSENTIALS

D1451267

AUTODESK® INVENTOR® 2013 AND AUTODESK® INVENTOR LT™ 2013

ESSENTIALS

Thom Tremblay

John Wiley & Sons, Inc.

Senior Acquisitions Editor: Willem Knibbe
Development Editor: Mary Ellen Schutz
Technical Editor: Dan Hunsucker
Production Editor: Rebecca Anderson
Copy Editor: Kim Wimpsett
Editorial Manager: Pete Gaughan
Production Manager: Tim Tate
Vice President and Executive Group Publisher: Richard Swadley
Vice President and Publisher: Neil Edde
Book Designer: Happenstance Type-O-Rama
Compositor: James D. Kramer, Happenstance Type-O-Rama
Proofreader: Jen Larsen, Word One New York
Indexer: Ted Laux
Project Coordinator, Cover: Katherine Crocker
Cover Designer: Ryan Sneed
Cover Image: Thom Tremblay

Copyright © 2012 by John Wiley & Sons, Inc., Indianapolis, Indiana

Published simultaneously in Canada

ISBN: 978-1-118-24479-1
ISBN: 978-1-118-33326-6 (ebk.)
ISBN: 978-1-118-33043-2 (ebk.)
ISBN: 978-1-118-33112-5 (ebk.)

For general information on our other products and services or to obtain technical support, please contact our Customer Care Department within the U.S. at (877) 762-2974, outside the U.S. at (317) 572-3993 or fax (317) 572-4002.

Wiley publishes in a variety of print and electronic formats and by print-on-demand. Some material included with standard print versions of this book may not be included in e-books or in print-on-demand. If this book refers to media such as a CD or DVD that is not included in the version you purchased, you may download this material at http://booksupport.wiley.com. For more information about Wiley products, visit www.wiley.com.

Library of Congress Control Number: 2012936407

10 9 8 7 6 5 4 3 2 1

Dear Reader,

Thank you for choosing *Autodesk® Inventor® 2013 and Autodesk® Inventor LT™ 2013 Essentials*. This book is part of a family of premium-quality Sybex books, all of which are written by outstanding authors who combine practical experience with a gift for teaching.

Sybex was founded in 1976. More than 30 years later, we're still committed to producing consistently exceptional books. With each of our titles, we're working hard to set a new standard for the industry. From the paper we print on, to the authors we work with, our goal is to bring you the best books available.

I hope you see all that reflected in these pages. I'd be very interested to hear your comments and get your feedback on how we're doing. Feel free to let me know what you think about this or any other Sybex book by sending me an email at nedde@wiley.com. If you think you've found a technical error in this book, please visit http://sybex.custhelp.com. Customer feedback is critical to our efforts at Sybex.

Best regards,

NEIL EDDE
Vice President and Publisher
Sybex, an Imprint of Wiley

To Nancy, for everything.

ACKNOWLEDGMENTS

I want to thank the tremendous team at Sybex for their patience and professionalism, specifically, Willem Knibbe, Pete Gaughan, Mary Ellen Schutz, Rebecca Anderson, Kim Wimpsett, and everyone else who worked hard behind the scenes who I didn't get a chance to communicate with directly. Special thanks once again to Dan Hunsucker for being the technical editor. If you're in the Kansas City area and want to learn to use the Autodesk® Inventor® software from a real expert, you'll be in great hands with Dan. Thanks to Tom Joseph, Sean James, and Matt Pierce of Autodesk for their support. Of course, many thanks to my family for putting up with my absence on the weekends and nights.

About the Author

Thom Tremblay is the Industry/University Relationship Manager on the Autodesk Strategic Universities team and has worked with hundreds of companies to help them understand how Inventor can help them with their designs. He is an Autodesk Inventor Certified Professional and has been working with Inventor for more than 10 years and with other Autodesk products for more than 25 years. He has used Autodesk software to design everything from cabinets and castings to ships and video monitors. He has close ties to the Inventor community; is a frequent speaker at colleges, universities, and training centers; and presents at Autodesk University annually.

Contents at a Glance

CONTENTS

CHAPTER 10   Working with Sheet Metal Parts   221

CHAPTER 11   Building with the Frame Generator   245

CHAPTER 12   **Working in a Weldment Environment   259**

CHAPTER 13   **Creating Images and Animation
from Your Design Data   273**

Introduction

Welcome to Autodesk Inventor 2013 and Autodesk Inventor LT 2013 Essentials. This book is intended to be a direct, hands-on guide to learning Inventor by using Inventor. The book includes lessons for absolute beginners, but experienced users can also find exercises to show them how tools they're not familiar with work.

Nearly all of the 200 exercises can be started from an existing file, so you need to do only those exercises that will help you most.

Who Should Read This Book

Autodesk Inventor 2013 and Inventor LT 2013 Essentials is designed to meet the needs of the following groups of users:

▶ Professionals who use 2D or 3D design systems and want to learn Inventor at their own pace

▶ Professionals attending instructor-led Inventor training at an Autodesk Authorized Training Center

▶ Engineering and design students who need to learn Inventor to support their education and career

What You Will Learn

Autodesk Inventor 2013 and Autodesk Inventor LT 2013 Essentials covers the most common uses of the tools in Autodesk Inventor and Inventor LT. Not every option is covered, but you will easily understand the options once you learn to use the primary tool.

The first eight chapters cover the core of Inventor in a way that steps the reader through creating drawings, parts, and assemblies in phases so there is a better opportunity to absorb the concepts.

The second portion of the book consists of seven chapters that focus on tools and workflows specific to types of design and on reusing data and automating the design process. I recommend readers complete the work in these chapters to learn alternative workflows that may not be an obvious fit for their design needs but may help them nonetheless.

What You Need

To perform the exercises in this book, you must have Autodesk Inventor 2013 or Inventor LT 2013.

While offering a somewhat limited features set, Inventor LT is a flexible and powerful tool focused on the core user needs of Part Modeling, 2D Drawings, and importing data from other sources.

Chapters 1–3, 5–7, and 14 focus on these core tools. You will also find that Inventor LT is capable of doing some of the functions covered in chapters 10 and 13. In cases where an assembly is used for an exercise on visualization you can use a part file instead.

To make sure that your computer is compatible with Autodesk Inventor 2013, check the latest hardware requirements at www.autodesk.com/inventor.

What Is Covered in This Book

Autodesk Inventor 2013 and Autodesk Inventor LT 2013 Essentials is organized to provide you with the knowledge needed to master the basics of Inventor.

Chapter 1: Connecting to Inventor's Interface   This chapter presents the interface, the basics of working with Inventor, and how to become productive with Inventor.

Chapter 2: Creating 2D Drawings from 3D Data   Creating 2D documentation of your designs is critical. This chapter presents the basic tools for placing views and dimensions in your drawings.

Chapter 3: Learning the Essentials of Part Modeling   Building parametric solid models is essential to the effective use of Inventor. This chapter will introduce the tools you need to build basic parts in Inventor.

Chapter 4: Putting Things in Place with Assemblies   Most products are made of many parts. Assembly tools help you control the position of the components relative to one another.

Chapter 5: Customizing Styles and Templates   Using standards in manufacturing improves quality and efficiency. The same is true for Inventor. This chapter helps you understand the options that are available for building your own design standard.

Chapter 6: Creating Advanced Drawings and Detailing   This chapter focuses on creating and editing more complex drawing views and adding finishing touches to your drawings.

Chapter 7: Advanced Part Modeling Features   Advanced geometry requires more advanced modeling tools. Learn the use of advanced fillets, lofts, and other tools that create the complex shapes you need.

Chapter 8: Advanced Assembly and Engineering Tools   An assembly is more than a group of parts. Inventor features many engineering-based tools that work in the assembly. This chapter also describes tools to help you control complex assemblies.

Chapter 9: Creating Plastic Parts   Plastics have a number of common features that make them easier to assemble. These features are developed using specialized tools in Inventor.

Chapter 10: Working with Sheet Metal Parts   The process of manufacturing sheet metal parts heavily influences how they are designed in Inventor. Creating material styles makes it easy for you to change the components by changing the style.

Chapter 11: Building with the Frame Generator   Using traditional solid modeling tools to build metal frames is arduous and time-consuming. The Frame Generator tools shortcut the process and make even complex frames easy to design.

Chapter 12: Working in a Weldment Environment   A weldment is a combination of an assembly and a part model. Inventor puts the needs of manufacturing first when defining a weldment, saving you time.

Chapter 13: Creating Images and Animation from Your Design Data   Sharing images and animations made from your designs can help others understand how your designs are created and understand their value. This chapter guides you through tools for sharing your work with others.

Chapter 14: Working with Non-Autodesk Inventor Data   Inventor has the ability to import and export data to and from nearly any other design system. This chapter helps you understand what options you have in working with that data.

Chapter 15: Automating the Design Process and Table-Driven Design   If you have repeatable design processes and products that share a lot of component families, this chapter can help you see opportunities to greatly increase your efficiency.

Appendix: Autodesk Inventor Certification   Show the world that you know Inventor by becoming an Autodesk Certified User, Associate, or Expert. This appendix will help you find the resources in the book to get certified.

Exercise Data   To complete the exercises in *Autodesk Inventor 2013 and Autodesk Inventor LT 2013 Essentials*, you must download the data files from `www.sybex.com/go/inventor2013essentials`.

Please also check the book's website for any updates to this book should the need arise. You can also contact the author directly by email at `inventor.essentials@yahoo.com`.

The Essentials Series

The Essentials series from Sybex provides outstanding instruction for readers who are just beginning to develop their professional skills. Every Essentials book includes these features:

- ▶ Skill-based instruction with chapters organized around projects rather than abstract concepts or subjects.

- ▶ Suggestions for additional exercises at the end of each chapter, where you can practice and extend your skills.

- ▶ Digital files (via download) so you can work through the project tutorials yourself. Please check the book's web page at `www.sybex.com/go/inventor2013essentials` for these companion downloads.

CHAPTER 1

Connecting to Inventor's Interface

To access the power of Autodesk Inventor 2013, you have to start with the interface. To some extent, Inventor is an interface between your ideas and your computer.

The ability to navigate and leverage the nuances of a program interface can be the difference between struggling and excelling with the application. In this chapter, you will see the components of Inventor's dialog boxes, Ribbons, tabs, and viewing tools that will help you create your designs. You will also learn how to modify the interface to increase your comfort with Inventor.

▶ **Exploring inventor's graphical user interface**

▶ **Setting application options**

▶ **Using visualization tools**

▶ **Working with project files**

Exploring Inventor's Graphical User Interface

When you first see Inventor's interface, you will probably think it is rather bare. With no file open, you just have the absolute basics there. Even when a file is loaded, your design remains the focus of the interface. In Figure 1.1, you can see the primary elements of the interface that will be referred to in this chapter.

Users of other current Autodesk or Microsoft applications will recognize the Ribbon-style interface and the Application icon in the upper left. Inventor's adoption of the Ribbon interface goes beyond most other applications by actively offering you tools when they're most needed. But let's not get ahead of ourselves; let's start by getting more detail on these features of the interface.

Application Menu   Quick Access Toolbar   Title Bar   Ribbon   Help Tools   ViewCube

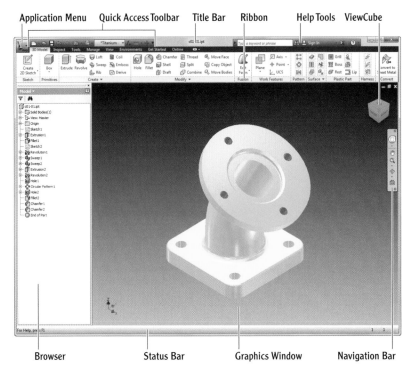

Browser   Status Bar   Graphics Window   Navigation Bar

FIGURE 1.1   Elements of the Inventor user interface

Across the top of the Inventor window is the title bar. It lets you know you're using Autodesk Inventor, or it displays the name of the active file when you're editing one.

In the upper-left corner is an icon with a large *I* on it. Clicking it opens the Application menu (Figure 1.2), which displays tools for creating and manipulating files (on the left) and a list of recently opened files (on the right). If you want to be able to return to a file frequently, you can select the pushpin icon to the right of the filename and keep it on the list of recently accessed files.

You can also toggle the list between recent documents and the documents that are currently open and change the list from filenames to icons showing the files.

At the top (on the right side) is a *Search Commands* window for finding tools that are available in the file you are editing. As you begin typing a tool name or more conceptual term a list of options based on the entry will be generated. You can then start the tool from that list. The location of the tool in the Ribbon is also displayed for future reference.

Just below that you will find icons that sort how previously opened files are listed and a drop-down menu for the size of the icon displayed.

FIGURE 1.2 The expanded Application menu and Quick Access toolbar

At the bottom of the menu are buttons to exit Inventor and to access the application options, which you will explore later in this chapter.

The Quick Access toolbar is embedded in the title bar next to the Application icon and contains common tools to access new file templates, undo and redo edits, and print. The toolbar is dynamic, and different tools will appear depending on the active file. For example, one of those part-time tools is a drop-down menu that allows you to change the color of the active part.

You can customize this toolbar by adding commonly used tools to the toolbar. To do this, select the desired tool from the Ribbon, right-click, and select Add To Quick Access Toolbar from the context menu.

Certification
Objective

IMPORTANT!

At this time, it is critical that you go to www.sybex.com/go/ inventor2013essentials, download the Inventor 2013 Essentials Data.zip file, and expand it to the root of your C drive. This will create a new folder, C:\Inventor 2013 Essentials. The files needed to complete the exercises for this book are in this folder.

Opening a File

Knowing this much of the interface will allow you to access the Open dialog box and see how the rest of the interface works. In this exercise, you will open a file in Autodesk Inventor:

1. Start Autodesk Inventor if it is not already up and running.

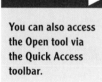

2. Expand the Application menu, and select Open from the options on the left.

   When the Open dialog box appears, notice the tools at the top. These tools allow you to navigate to other folders as you would in Windows Explorer, to change the way files are displayed (including the option of thumbnail images), and to add new folders.

> **You can also access the Open tool via the Quick Access toolbar.**

3. In the Open dialog box, use the Browse button to navigate to the `C:\ Inventor 2013 Essentials\Parts\Chapter 1`, folder, as shown in Figure 1.3.

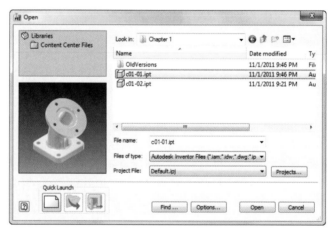

FIGURE 1.3 The Open dialog box includes tools for finding and browsing for files.

> **Selecting an Inventor file from the file list generates an image in the Preview pane on the left side of the dialog box.**

4. Double-click the `c01-01.ipt` file listed, or click it once and then click Open.

5. To see the complete model, move your cursor near the ViewCube in the upper right of the Graphics window. When the icon that looks like a house appears, click it.

Clicking the Home view icon restores a view position that was saved with the model. Your screen should now resemble the earlier Figure 1.1.

Certification
Objective

KNOW YOUR ICONS!

These are the icons for the four primary Inventor file types:

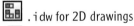 .idw for 2D drawings

.ipt for 3D parts

.iam for 3D assemblies

.ipn for 3D presentation files

The Open dialog box, templates, and even Windows Explorer use these icons as a visual reference for the type of information they contain.

Exploring the Ribbon

Certification
Objective

You will now see that the Ribbon has more options available. These options take the form of tools that are grouped into panels on tabs. Let's look at these in order from the broadest to the most specific:

Tabs　The options shown as 3D Model, Inspect, Tools, and so on, are referred to as *tabs*. The active tab always contrasts with the others that just appear as names on the background of the Ribbon. The active tabs can change automatically as you transition from one working environment to another.

There are tabs that appear temporarily when you use specialized tools or enter a specialized environment, such as sketching (Figure 1.4) or rendering a 3D model. These *contextual tabs* activate automatically and have a special green highlight to help them stand out and remind you that the tools are available.

FIGURE 1.4　The Sketch tab is not normally displayed but gets special highlighting when it is.

Panels　On each tab are special collections of tools that are sorted into *panels*. For example, in Figure 1.4, you can see Draw, Constrain, Pattern, and other panels on the Sketch tab. Notice that the most commonly used tools appear with larger icons so they're easier to locate. At times, not all of the tools in a panel can fit, so some of the panels (Draw and Constrain) have down arrows next to their names. This indicates that you can expand the panel to see more tools. It is also possible to select tools that you don't often use and place them permanently in this expandable portion of the panel. You can even rearrange the order of the panels in a tab.

> Any time you see a down arrow next to a word or icon in Inventor, it means there are additional options.

Tools The last element of the Ribbon is the icon or tool. Many of the tool names display near the icon. As you become more comfortable, you can save space on your screen by turning off the display of the names. You can also hide or partially hide the Ribbon if you prefer having a larger Graphics window.

Let's try a few of these options so that you can make yourself more comfortable once you start using Inventor for your regular work:

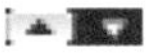

1. Keep using or reopen the `c01-01.ipt` file used in the previous exercise.

   To the right of the tab headers is a gray icon with an up arrow. Click this arrow to cycle the various ways of displaying the Ribbon.

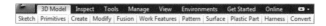

2. Click the icon once to change the Ribbon from displaying all the panels within a tab to showing only the first icon with the title of the panel below it, as in Figure 1.5.

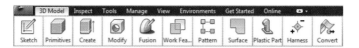

FIGURE 1.5 The 3D Model tab showing Panel Buttons mode

   When you hover over or click the icons that represent a panel (or the panel or tab titles of the next steps), a full view of the panel appears under the Ribbon.

3. Click the same icon a second time, and you will reduce the tab to showing only the title of the panel, as in Figure 1.6.

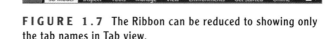

FIGURE 1.6 Another cycling of the Ribbon viewing options shows the Panel Title mode.

4. Now, click the icon again to reduce the Ribbon even further, as you can see in Figure 1.7.

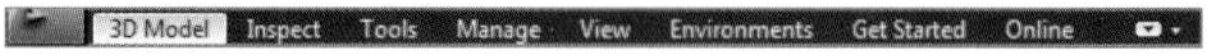

FIGURE 1.7 The Ribbon can be reduced to showing only the tab names in Tab view.

5. Click the icon one last time to restore the Ribbon to its original size with the default panels displayed.

Depending on your needs and personal preferences, you can use these various modes full-time or just temporarily. Clicking the down arrow to the right of the icon displays a menu that allows you to jump directly to the mode you want.

Rearranging the Panels

Let's do one more exercise displaying the Ribbon tools.

1. Keep using or reopen the c01-01.ipt file used in the previous exercise.

2. Click the Model tab.
    The title bar beneath the panel can be used as a grip.

3. Click and drag the Pattern panel (Figure 1.8) so that it appears between the Create and Modify panels.

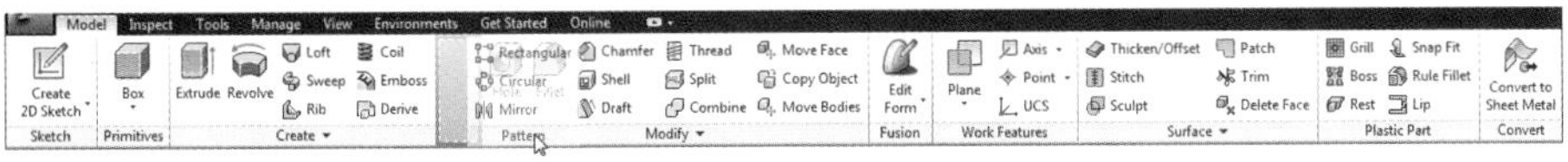

FIGURE 1.8 Any panel can be relocated to suit your tastes.

4. Right-click one of the panels, and expand the Ribbon Appearance menu by hovering over the words in the context menu.

5. Select Text Off from the menu.
    This will remove the names of the tools in the panel. This cuts down on the screen area covered but keeps the tools readily available.

6. Right-click again in a panel to open a menu with Ribbon appearance options.

7. Select Reset Ribbon.

8. Click Yes when the warning dialog box appears.

After a few moments, the icon names reappear, and the panels revert to their original positions. This is a great safety net for experimenting with interface personalization.

Finally, note that hovering your cursor over a tool in the Ribbon will open a *tooltip* that will give you basic information, the correct name of the tool, and, when applicable, the keyboard shortcut for the tool. If you leave the cursor over that tool a little longer, the tooltip will expand to offer visual information about the use of the tool (Figure 1.9). Some tooltips even offer animated guidance on the tool.

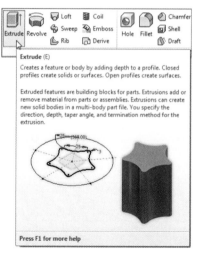

FIGURE 1.9 Continuing to hover expands the tooltip.

Using the Browser

Inventor's process of design is often referred to as *parametric solid modeling*. Although parameters play an important role in using Inventor, what makes its design process powerful and flexible is its ability to maintain a history of how the parts were constructed. This history also shows how they are related to one another in an assembly. The interface to see these actions is the browser.

On the left side of the screen is a column that contains a hierarchical tree showing the name of the active file at the top, the Representations and Origin folders, and then the components or pattern of components that make up your assembly. Understanding how to read the browser and how to control it is an important step in understanding how to control and edit your designs.

Let's do some exploration of the browser to see what it can tell you:

1. Keep using or reopen the c01-01.ipt file used in the previous exercise.

2. In the browser, find the Origin folder, and click the plus sign to the left of the folder to expand the folder's contents.

3. Pass your mouse over the planes and axes displayed in the browser, and see how those entities preview in the Graphics window to the right.

4. Right-click the XY Plane icon, and select Visibility from the context menu that appears.

The browser displays the features that make up the active part (see Figure 1.10). Since it is a part file, the Ribbon has also changed, making the Model tab active for adding features to the part.

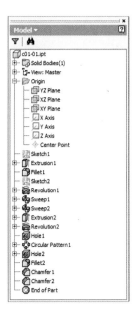

FIGURE 1.10 The browser reflects the status of the part or assembly and its history.

If you activate a part in an assembly, it's also important to know how to return to the assembly view. On the right end of the Model tab, you would see the Return panel. Click the icon for the Return tool. This tool will appear at the end of a number of Ribbons when editing parts and assemblies.

In an assembly, there are additional options in the Return tool that you will use later in the book. Return's primary function is to take you up a step in the hierarchy of the model you're in.

Not only is Inventor capable of creating and editing parts while in the assembly, but it prefers to work that way. There is no need for Inventor's sake to open a part in a separate window to edit it, though at times you might find it easier to navigate. Once a part is activated, the inactive components in the assembly are grayed out in the browser and fade in the Graphics window to make it easier to focus on the part you're editing.

Exploring the File Tabs

Once you've opened at least two files in Inventor, the File tabs appear, to make it easier to navigate between the files, as well as to find the file that you want to switch to.

Let's open another part to see what other changes occur when you open a separate window to use for editing. Open the C:\Inventor 2013 Essentials\ Parts\Chapter 1\c01-02.ipt file in Inventor. In Figure 1.11, you can see the

If you're not using Inventor LT, you can also activate components in an assembly by double-clicking the geometry in the browser or the Graphics window. Selecting a component and then right-clicking will allow you to open the part in its own window.

file tabs at the bottom of the Graphics window. These tabs display the names of the files.

FIGURE 1.11 File tabs let you quickly switch between open files and even preview the file before you select it.

The first two buttons on the left arrange your open files by cascading them or by displaying them as tiles side by side. Let's try using the tools to see what happens:

1. Click the Arrange button on the file tabs bar.

2. The files will stack on top of each other, as shown in Figure 1.12.

3. Look at the browser, and then click back and forth between the tabs for the c01-01.ipt and c01-02.ipt files to see the effect on the browser.

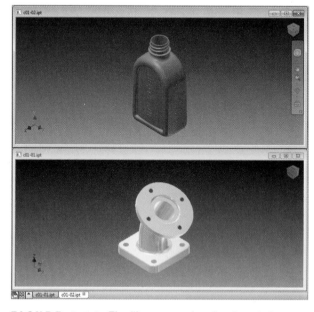

FIGURE 1.12 The files arranged as floating windows

4. Select the c01-02.ipt tab, and click the Close button (small X icon) in the upper right of the tab.

   This closes the c01-02 part and the file tabs but leaves the other part open in a floating window.

5. Double-click the top of the floating window frame, or click the Maximize button near the upper-right corner of the window.

This will restore the c01-01 part to filling the Graphics window. The Graphics window has many different tools and enables many different options. I'll cover some of them in future chapters, but let's look at some of the most important ones now.

Highlighting and Enabled Components

In a part file, moving your cursor over faces and edges will highlight them for selection. The same is true for the views of a drawing. As you move your cursor over the components of an assembly in the Graphics window, you will notice that the parts highlight. Highlighting allows you to see when components stand alone or, as in the case of bolts in an assembly, as a group. When a component is highlighted in the Graphics window, it will also highlight in the browser, and vice versa.

Sometimes, you want to visually reference an assembly component but do not want to be able to select that component or activate it. To do this, you can select the part in the browser or the Graphics window and deselect Enabled from the context menu. This will fade the part, make it unselectable, and disable highlighting of the disabled part.

Working in the Graphics Window

The part of the interface where most of your attention will be focused is the Graphics window. Not only is your design and its documentation shown there, but with each release of Inventor, more tools can be started from the Graphics window. Viewing and visualization tools can affect this area and allow you to view and share your designs in exciting ways.

Certification Objective

Let's go through some of the tools that exist in and for the Graphics window and begin to see just how important it is:

Origin 3D Indicator　In the lower-left corner of the Graphics window, you see a 3D direction indicator. This is a reference for the orientation of the model in relationship to the standard axes of the file.

Pan　To slide the model or drawing around the Graphics window, use the Pan tool. You can access Pan in several ways. You can find the Pan tool by going to the Navigate panel on the View tab, by going to the Navigation bar in the Graphics window, by pressing the F2 key, or by clicking and holding the middle mouse button.

 Zoom   The Zoom tool has several optional modes. The basic Zoom tool smoothly makes the model appear larger or smaller depending on the input you give it. You can find the Zoom tool by going to the drop-down under the Navigate panel on the View tab, by locating the Zoom icon on the Navigation bar, by pressing the F3 key and clicking and dragging your mouse button, or by rotating a mouse wheel.

 Free Orbit   When you start Free Orbit by going to the Navigate panel, by going to Navigation bar, or by pressing F4, a reticle appears on the screen. Moving your cursor around the reticle displays different tools. Figure 1.13 shows icons inside, outside, and near the x- and y-axes of the reticle. Clicking inside the reticle orbits the model in a tumbling manner. Clicking and dragging near the x- or y-axis or just outside of the reticle isolates the orbit around an axis.

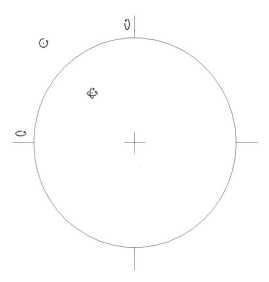

FIGURE 1.13 It is important to be able to easily reposition your view of the model for editing.

View History   You can return to a previous view or views by pressing and holding the F5 key until a view that you had before is restored on the screen.

The ViewCube   It's very likely the current view of the assembly isn't the only useful view that you might need to develop the design. The ViewCube allows you to rotate your design and position it based on standard views aligned with the faces, edges, and corners of the cube. Rotating the model using the ViewCube works in a way similar to the Free Orbit tool.

The ViewCube is probably the most important tool in the Graphics window for keeping track of your design. In the next exercise, you will get a chance to begin using it.

Certification Objective

1. Keep using or reopen the c01-01.ipt file used in the previous exercise.

2. Move your cursor to the ViewCube and over the face labeled Front.

3. When the face highlights, click it to rotate your model.

4. Now, click the arrow that is pointing at the right side of the cube. This rotates your view to the Right side.

5. Move your cursor to the upper-right corner of the ViewCube, and select it to see the Top, Right, and Back faces of the ViewCube.

6. Now, click the ViewCube, hold your mouse button down, and move your mouse around to orbit the view.

7. After you've rotated your part, press the F6 key to return to the Home view of the assembly.
   Now, your screen should look as it did at the beginning of the exercise.

8. Click the face labeled Right.

9. After the view updates, right-click the ViewCube, hover over Set Current View As, and select Front from the menu that appears (see Figure 1.14).

10. Now orbit the view to something you think offers a better view than the current Home view.

11. Right-click the ViewCube and use the Set Current View As Home with the Fit To View option.

FIGURE 1.14 Faces on the ViewCube can be redefined as the Front or Top.

The other option for defining a new Home view is Fixed Distance. This option maintains the zoom scale and position, even if additions to the model go outside the Graphics window.

The ability to define the orientation of the ViewCube and the Home view will prove useful as you gain experience.

Checking Out the Status Bar

On the bottom edge of the Inventor window is a bar that gives you information about expected inputs for tools. The status bar also gives you feedback on information needed by Inventor, as well as the number of files that are being accessed and parts that are being displayed when working in an assembly.

Using Marking Menus

As you create drawing views, parts, and assemblies and as you use other tools, you will see dialog boxes for these tools.

Beginning with Inventor 2011, several "Heads-Up" interface tools have been added that put common commands at the mouse cursor. Marking menus work in two modes: Menu and Marking.

Menu Mode   When you are editing a file and right-click in the Graphics window, a "compass" (see Figure 1.15) of tools appears. As you move toward a tool, a shadow will highlight it, and clicking the mouse with this highlighting active selects the tool.

In the Options panel of the Tools tab is a Customize tool. With it, you can change what tools appear on the compass and how they look. You can also control the size of the Overflow menu that appears beneath the compass.

FIGURE 1.15 Marking menus place common tools where they are needed most.

Marking Mode   Once you're experienced with the tools offered in the marking menu, you can begin using Marking mode. To do this, right-click and drag the cursor toward the tool you want to select. When you release the button, you will activate the tool. The tool will flash to confirm that you've selected it rather than showing all the options.

Setting Application Options

Many settings allow you to tailor Inventor to your needs. Changing the tools in the Ribbon is just the beginning. To really affect the way Inventor behaves, you will need to make changes to the application options.

Certification Objective

To access the Application Options dialog box, you can go to the Application menu and click the Options button at the bottom; or, go to the Tools tab and select the Application Options tool. This will open a large dialog box with several tabs and an enormous number of options. I will focus on a few of these options, but I strongly recommend that you explore the others.

Once you are in the Application Options dialog box, you will see that the options are categorized in several tabs. I will suggest a few key settings for some of those tabs in the following sections.

Using the Import/Export Buttons

At the bottom of the Application Options dialog box are buttons that allow you to export your settings for backup or to share with others. You can also import one of these backups or import the settings another user has.

Exploring the General Tab

Once you're comfortable with how you typically use Inventor, you might want to make changes to some of the basic settings. On the General tab, you can change how Inventor starts up, how long you have to hover over an icon before the tooltip will appear, and even how much room on your hard disk can be used by the Undo tool.

One option on this tab is to automatically update the physical properties of the model when you save. This will keep things like the annotations added to the title block in Chapter 5, "Customizing Styles and Templates," up-to-date without an extra step by the user.

Exploring the Colors Tab

The Colors tab has the favorite tools for exploration for most Inventor users. You can change the color and look of your Graphics window background, control the color of the sketch objects and dimensions, and change the color scheme of the Ribbon icons.

Follow these steps to save your current settings and experiment with some changes:

1. Keep using or reopen the c01-01.ipt file used in previous exercises.

2. Expand the Application menu, and click the Options icon.

3. At the bottom of the dialog box, click the Export button.

A dialog box appears showing a couple of default settings backups. You can save your settings in this folder after you've made some changes.

4. Click Cancel.

5. Now, select the Colors tab.

6. Select Presentation from the Color Scheme list, and notice the change in the preview window.

7. Drag the title bar of the dialog box to the left so you can see part of the Graphics window of Inventor behind.

8. Click the Apply button at the bottom of the dialog box. See Figure 1.16 for reference.

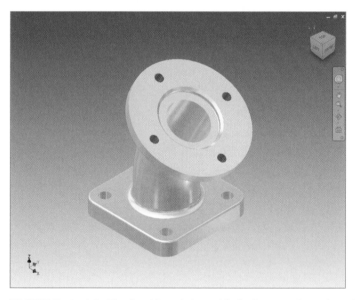

FIGURE 1.16 The Graphics window with the Presentation color scheme

9. Now, use the Background drop-down list, and change the setting to 1 Color.

10. Click Apply to see the effect.

Now try experimenting with different Color Scheme values, as well as changing the color theme and background, until you find something that is pleasing to you. You can also use an image file for the background of your screen to give additional realism or create a special effect.

For this book, I will use the Sky Color scheme with a white background image to keep the components in the Graphics window easy to see for the printed graphics. Using the Presentation scheme with a 1 Color background is a quick way to generate images for your technical documents as well.

Exploring the Display Tab

The Display tab controls a number of display effects you've already encountered. On the Display tab, you can use the Inactive Component Appearance section to control the fading effect on inactive parts while editing a part in the assembly.

Similarly, when you selected the Home view, the model had an animated transition into position. The speed of that transition is controlled with the slider bar under Display.

Changing the Minimum Frame Rate (Hz) value to a higher number causes Inventor to make parts of an assembly invisible while zooming, panning, or orbiting in order to maintain the chosen frame rate. The real use of this value is to allow you to change the view quickly while still being able to see what position the model is in.

Other settings allow you to reposition the ViewCube on the screen and control the visibility of the Origin 3D indicator or its axis labels.

Exploring the Hardware Tab

Most of the settings for hardware are fairly self-explanatory. A very important item is the Software Graphics check box. If you experience frequent crashing in Inventor, you should select this check box (see Figure 1.17), restart, and see whether the crashing stops.

FIGURE 1.17 Selecting the Software Graphics option can help overcome stability issues.

Inventor's Help system uses multiple resources to locate information for you. The Help tools are on the right end of the title bar, and they include a search window that will locate information from multiple resources, including the wiki for Inventor. You can even share your own tips on the wiki.

The majority of crashes in Inventor begin as a conflict between the graphics driver, Windows, and Inventor. Selecting Software Graphics disconnects Inventor from communicating directly with the graphics driver. If the crashing stops, you should verify that your video card is supported and make sure you have the certified driver for the card. The Additional Resources section under Help has a link to look up the certified driver for approved video cards.

The Diagnostics tool at the bottom can help test whether the graphic drivers are WHQL-certified for Windows and whether your graphics card is capable of offering acceleration.

Exploring the Assembly Tab

This tab's settings are primarily at-large assembly needs and specialized options. You can control whether the inactive components of the assembly remain opaque when a component is activated for editing; you can also change this option on the View tab by extending the Appearance panel.

If you decide that you would rather not hear the sound played when you place an assembly constraint, you can turn it off on this tab.

Exploring the Drawing Tab

If you choose to create drawings in DWG format, they can be created in 2000, 2004, 2007, 2010, or 2013 format.

This tab has some settings that should be very interesting to AutoCAD users. The Default Drawing File Type setting can be selected, which allows you to create drawings in the traditional Inventor IDW format, or you can choose to create DWG files that can be opened for viewing, printing, and measurement in AutoCAD. You can even set which AutoCAD release format a DWG will be created in.

You can set some overall dimension preferences, and you can control whether line weights are displayed. For users with limited graphics power, you can change the View Preview Display setting to see views with a full preview, with a partial view, or with a bounding box.

Exploring the Sketch Tab

In Chapter 3, "Learning the Essentials of Part Modeling," you will begin exploring part modeling. When you first begin to use sketching tools, a grid can be displayed. For clarity, the graphics in this book will not display the grid, minor grid lines, or axes of the sketch. If you prefer not to see these, you can disable parts or all of the grid in the Display section of the Sketch tab.

In the Heads-Up Display settings, you can control how dimensions are displayed on entities as you sketch them. Many people prefer to disable the tools that automatically project edges during sketch or curve creation. These tools typically include Autoproject in the description. Many users also choose not to look at the sketch plane when sketches are created. You can adjust for those preferences here.

Exploring the Part Tab

A common change made on the Part tab is to enable a sketch creation when new parts are created. You can also choose to automatically have sketches created on a specific plane when you start a new part.

If you would like to begin new parts by sketching right away, you can set the plane you would like to begin on in the *Sketch On New Part Create* frame.

This tab can also control the editing tool used to modify imported geometry between Inventor's legacy tools and Autodesk Inventor Fusion. You will get a chance to explore Inventor Fusion in Chapter 14, "Working with Non-Inventor Data."

Using Visualization Tools

The ability to communicate a design visually is a great asset of Autodesk Inventor. Changing the color of the Graphics window or setting the background to be an image can give an interesting effect.

Increasingly, designers use renderings to present new concepts rather than creating detail drawings that might not be understood by people setting budgets. Regardless of the reason, the visualization capabilities that reside within the modeling environment can be just plain fun.

Understanding the Visual Styles

 Certification Objective

You can find the tools covered in this section on the View tab in the Appearance panel. I will review many of them, but I'm sure you will find yourself experimenting with them for a long time to come.

Realistic   The Realistic visual style works at two levels. First, it uses the common color library that is shared among most Autodesk design products, giving you a different appearance than the other modes. Realistic also lets you very quickly create a ray-traced rendering that is more than suitable for communicating a design concept. Although this option is in the modeling environment, it really isn't intended to be the setting used while doing modeling or assembly functions. Not all graphics cards are capable of working with the ray-tracing capabilities.

Shaded   Shaded mode has been available throughout Inventor's history and is the environment that most users work in.

Shaded With Hidden Edges   To highlight the visible edges of the components and generate hidden edges that show through the components to make it easier to find faces that are otherwise obscured, use Shaded With Hidden Edges.

See Figure 1.18 for an illustration of these three modes.

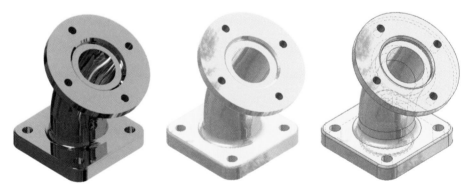

FIGURE 1.18 The Realistic, Shaded, and Shaded With Hidden Edges visual styles

Wireframe With Hidden Edges   Working in wireframe is comfortable for users who have experienced older 3D systems. The addition of displaying the hidden edges makes it easier to sort out what is in the foreground and in the background.

Monochrome   Monochrome makes your model look like Shaded mode but with grayscale shades replacing the colors of the components. It is a useful tool to see how a noncolor print of your shaded model might look.

Illustration   Illustration mode is in some ways a mirror to Realistic. It makes the model appear more like a line drawing by adding edges and varying line weights to the display.

See Figure 1.19 for the last three modes described.

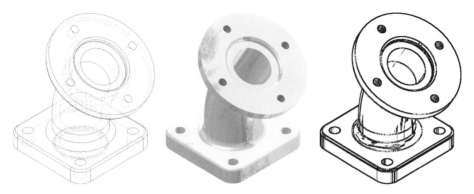

FIGURE 1.19 Wireframe With Hidden Edges, Monochrome, and Illustration visual styles

Using Shadows

Shadows can assist in understanding the contours of a part, in particular when a part is prismatic and faces can be obscured in Shaded mode. These shadows can be displayed all at once, in any combination, or individually.

Ground Shadows   An easy way to add a visual appeal to your model is to have it cast a shadow on the ground. The position of the ground is controlled by the location of the Top face on the ViewCube. The effect is of light shining down on the top of the model.

Object Shadows   Selecting the Object Shadows option causes components and features to cast shadows on one another based on the direction of the lighting.

Ambient Shadows   Ambient Shadows mode gives a somewhat more subtle but very effective change to the model but adds the effect of ambient occlusion, which causes areas where sharp edges meet to be darkened. See Figure 1.20.

At the end of the Shadows flyout menu is a Setting button. When selected, it brings up a dialog box where you can establish the direction from which the shadows will be cast. You can also control the density and softness settings, as well as the intensity of ambient shadows. You can also turn on image-based lighting here; you will look at that effect a little later in the chapter.

FIGURE 1.20 With the model in Shaded mode, turning on Ambient Shadows gives it a little more realism.

Using Ground Reflections

 Ground reflections are based on a calculated "ground" that lies beneath the lowest component feature or visible sketch point. The ground establishes its location based on the position of the Top face of the ViewCube. The Ground Reflections option is either on or off, but the fun doesn't have to stop there.

Like the Shadows tool, Ground Reflections provides settings in a flyout menu next to the icon. In the settings, you can control the level of reflection, how much it is blurred, and how quickly that blur falls off. This dialog box also controls the settings for the Ground Plane tool, which will be described next.

Using the Ground Plane

This option is really just a switch that controls whether a grid will be displayed on the ground that the reflections and shadows are cast on. As mentioned, its settings are shared with Ground Reflections but can be accessed from its own flyout menu.

Putting Visual Styles to Work

In this exercise, you will get a chance to explore a couple of the options just reviewed:

1. Keep using or reopen the c01-01.ipt file used in previous exercises.

2. Switch your Ribbon to the View tab.

3. In the Appearance panel, click the icon for Shadows, and observe the change to the screen.

4. Right below the Shadows tool, click the Ground Reflections icon.

5. Next locate the Ground Plane tool, and turn it on.
   You will now use some of the tools that were discussed earlier.

6. Use the Visual Style drop-down menu to set your style to Realistic.

7. Now change your Visual Style setting to Watercolor. Note that the reflections and ground shadows disappear but ambient shadows are still generated.

8. Set Visual Style to Shaded when you're done admiring the Watercolor style, and keep this file open.

With a nice basic view, let's make a couple of simple changes to see what the effects are before looking at a few more visualization options:

1. Open the Shadows settings dialog box by using the flyout menu next to the Shadows icon.
   If possible, reposition the dialog box to see the screen as you make the following changes (because the screen will update while you are making them).

2. At the bottom of the dialog box, use the drop-down menu to change Shadow Direction to Light1.

3. Use the slider bar below to change the Density value to 40.

4. Change the Softness value to 80.

5. Reduce Ambient Shadows to 40.

6. Click the Save button and then Done to close the dialog box. See Figure 1.21, and keep the file open for the next exercise.

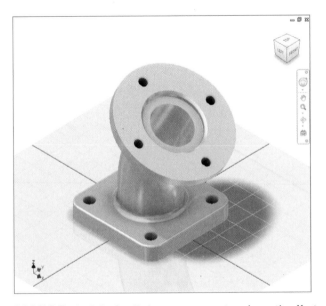

FIGURE 1.21 Small changes can create a dramatic effect with visualization settings.

With your shadows updated, you can do a little experiment with the reflections:

1. Open the Ground Reflections settings dialog box using the flyout menu next to the Ground Reflections icon.

   If possible, reposition the dialog box to see the screen as you make the following changes (because the screen will update while you are making them).

2. Set the Reflection level to 50%.

3. Set the Blur level to 50%.

4. Now change the value in the Minor text box under Appearance to 1.

5. Remove the check mark from Plane Color.

6. Set the Height Offset value at the top of the dialog box to 13mm.

7. Click OK to save the changes and close the dialog box.

8. In the Appearance panel, find the drop-down menu for orthographic and perspective projections, and set the view to Perspective.

Another option, Perspective With Ortho Faces, can be turned on by right-clicking the ViewCube and selecting it.

9. Press the F6 key to restore the Home view, and compare your screen to Figure 1.22. Keep the file open for the next exercise.

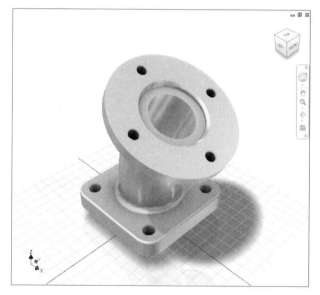

FIGURE 1.22 Some changes might decrease realism but can add energy to the view.

Remember, these changes aren't just for a printed document; you're still in a working part file, and you are able to make edits with these view characteristics.

Setting the Lighting Style

Lighting Style is a very understated name for a setting. Lighting styles do, in fact, change how lights (and therefore shadows) are positioned in your model, but several of them also come with an environment that offers the effect of putting the model into a different world.

Through these settings, you can make the same changes to the lighting style that you were able to make to shadows, but you will be limited in the built-in selections of image lighting sources. This tool's effect is easier seen than described, so let's use it.

You will do only a couple of changes, or you may not get any further in the book. Some of these next changes might not work with all graphics cards.

1. Continue working in the open file.

2. In the Appearance panel, the Lighting Style drop-down currently shows the Two Lights style. Change its value to Old Warehouse.

3. Use the ViewCube or Orbit tool to rotate your part around and see the environment that is shown. Using Pan and Zoom can help with finding some interesting points of view too.

4. Change your lighting style to Empty Lab.

Setting a lighting style can really change a lot of things about how you look at and work with your model. It's also a lot of fun.

Using Color Override

When you begin to create 3D part models, you will use specific materials so that you can monitor and predict the weight and inertial properties of the components you are building. Materials have their colors assigned to them, but you are not limited to having steel look like steel, for example. Using the Color Override option, you can change its color to make it look like anything you want.

Inventor 2013 offers more than one library of colors and materials. The Inventor Material library includes colors that have been used for years by Inventor. There is also an Autodesk Appearance Library and an Autodesk Material Library. Using colors from the Autodesk libraries give you consistency in the appearance of the parts if you open them in 3DS Max or other Autodesk applications.

Changing a component's color is straightforward, but keep in mind that changing a component's color in the assembly does not change it in the source file. If you want to change the color of something everywhere, you need to do it in the part file.

How easy is it? Let's find out:

1. Continue working in the open file.

2. Find the Color Override drop-down in the Quick Access toolbar and select a different color.

    At the bottom of the Color Override drop-down you can change the library used for color selection.

> If the color you select has a texture, it will make the faces of the part look like there are additional features or real textures on them. To maintain graphics performance, you can switch the display of textures on and off in the Appearance panel of the View tab.

3. Change the library to the Autodesk Appearance Library, and set the color to Chrome – Polished Blue.

4. As a nice finishing touch, change your visual style to Realistic. See my results in Figure 1.23.

You can close the c01-01.ipt file by clicking the X icon in the upper-right corner of the Graphics window or through the Application menu. (There is no need to save the results of this exercise.)

One of the fundamental steps in learning the tools of Inventor is to know where your files are. For that, you need a project file. The project file can also predetermine several options for accessing libraries, including the appearance and materials libraries. Now you will see how to create one.

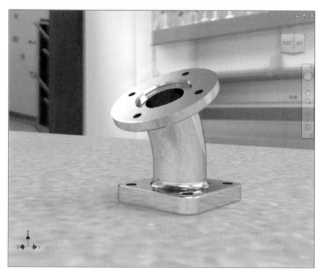

FIGURE 1.23 There is nothing some form of Chrome can't improve.

Working with Project Files

To work with Inventor effectively, you must have an understanding of the role of the *project file*. The project file is a configuration file that tells Inventor where to look for and store data about a project. Among other things, the project file can also be used to create shortcuts for locating files, change where Inventor looks for templates, and control the user's ability to modify standards.

It's also important if you work with others to have a consensus on how you will work with data and consistency in your project files. Working from the same

Certification
Objective

project file can save a lot of frustration. This can easily be done by simply copying the file among several users.

By now, you should've extracted the data for the files you will use in the book. Now, you will tell Inventor where to find that data.

Creating a Project File

You can create a project file at any time, but you can make it active only when all data is closed in Inventor.

1. Make sure that all files are closed in Inventor, but Inventor is still running.

2. On the Get Started tab, click the Projects icon in the Launch panel. This opens the Projects dialog box, where you can control the active project file or edit existing ones.

3. Click the New button at the bottom of the dialog box. This launches the Inventor Project Wizard, which will walk you through the process the next time you want to create a new project file.

4. Click the New Single User Project option, and then click the Next button at the bottom of the dialog box.

5. Enter 2013 Essentials in the Name field.

6. Enter C:\Inventor 2013 Essentials in the Project (Workspace) Folder field.

7. Compare your settings to Figure 1.24, and when you're sure everything matches, click Finish to create your new project file.

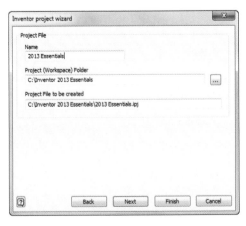

FIGURE 1.24 Defining the name and location of the new project file

Now you have a 2013 `Essentials.IPJ` file in the folder that you specified. (The `.ipj` extension stands for Inventor Project File.) You need to add some detail to make sure everything will be controlled as needed.

Modifying the Project File

In this exercise, you will add shortcut paths to folders that you will frequently access and make sure that the standard parts you use are kept where you need them.

1. In the top window of the Projects dialog box, double-click the 2013 Essentials project file to make it active.

2. In the bottom window, right-click Use Style Library, and switch it to Read-Write in the context menu.

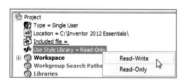

3. Click the Frequently Used Subfolders icon to highlight the row.

4. On the right, click the Add New Path icon.

5. A new row will be added under the row. In the field on the left, type **Assemblies.** On the right, browse to set your path to `C:\Inventor 2013 Essentials\Assemblies`. See Figure 1.25.

6. Click OK to set the path for the shortcut to the assembly files.
   When the new shortcut is created, you will see that the path is different from what it was in the dialog box. Instead of showing the full name, it will say *Workspace\Assemblies*. This helps avoid problems if the files need to be moved to another drive.

7. Add a shortcut for parts pointing toward the Parts folder and for drawings pointing to the Drawings folder.

8. Expand the Folder Options section.

9. Click the Content Center Files icon, and click the Edit icon on the right.

10. Set the path for the Content Center files to `C:\Inventor 2013\Content Center Files`, as shown in Figure 1.26, and then click OK to complete the edit. When prompted, save the changes.

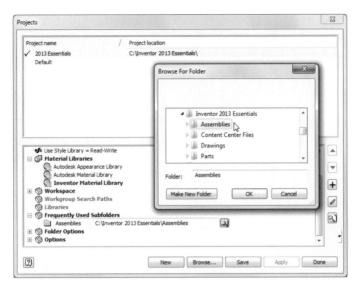

FIGURE 1.25 Creating an easy-to-understand alias representing a file path

FIGURE 1.26 Setting the path for saving standard components used in an assembly

As mentioned, the project file is critical for working with other users on the same data. The Content Center files path can be set to a common network location so that everyone can find the correct information when editing each other's work.

THE ESSENTIALS AND BEYOND

As you get more experience with the Inventor interface, you will see that it assists you in understanding which environment you are in and what is available to you. Changing colors and setting options that you like is a great way to get more comfortable, and that will lead to you being more productive.

(Continues)

THE ESSENTIALS AND BEYOND *(Continued)*

ADDITIONAL EXERCISES

▶ Add the tools that you think you will find useful to your Quick Access toolbar.

▶ Challenge yourself to access tools from different locations to become familiar with other tools around them.

▶ Use different application options, especially for sketching and part modeling, to see the effect they have on your productivity.

▶ Experiment with color schemes and lighting styles to find ways of maximizing the impression you make on people who see your work.

CHAPTER 2

Creating 2D Drawings from 3D Data

When you've never used Autodesk Inventor 2013, it is easy to focus entirely on the three-dimensional (3D) design tools, but many people still need to produce two-dimensional (2D) production drawings.

Chances are, if you're reading this book, you have created detail drawings and will likely need to create them in the future. Creating detail drawings from the solid (or surface) model is so easy it can often be thought of as fun.

To create a new drawing in Inventor, it is not necessary to have any 3D file open. In fact, if your system resources are limited, it is best not to have the file open.

▶ **Drawing views of a part**

▶ **Editing views**

▶ **Adding detail to drawing views**

▶ **Dimensioning**

Drawing Views of a Part

When you've created a drawing view, it must accurately reflect the model geometry, and if there is a change to the geometry, it needs to be reflected in all drawing views. This is just standard drafting practice, and the way Inventor works as well, but you may notice things happen a bit more quickly than on the drawing board.

This section introduces the types of drawing views and then proceeds through a series of exercises where you will create and work with views.

Types of Drawing Views

Any type of drawing view that you've ever created on the drafting board or in a 2D CAD system can be created with Inventor. It's just a lot faster in

Inventor because you're generating views of the object(s) rather than generating a lot of separate pieces of geometry to represent the same edge or feature. The following are the types of drawing views in Inventor:

The Base View   If you think about how you currently do drawing layouts, you are already working with a hierarchy, whether it's conscious or not. You have a view that others are positioned around. In Inventor, this is referred to as the *base view*, and there can be as many as you need or as many components as you like, all on one page.

Projected Views   Once a base view is placed, you can immediately place standard orthographic projections and isometric views from that base view. Any change that is made to the scale, position, or contents of the base view will be reflected in the projected view by default.

Section Views   These are placed much like projected views but with a section line defining what geometry will be generated. The section line can have multiple segments or include arcs.

> A section view can be made to any depth or even defined outside a part to create a custom view.

Auxiliary View   A specialized view requires the selection of an existing drawing view edge to project around to display the true shape of a face.

Detail View   This is another type of view that is hard to make in a 2D system, especially if the geometry may change. For this view, you can use round or rectangular boundaries with smooth or rough edges for the view. The detail view can also be created with any scale.

Starting a New Drawing

Certification Objective

To have a view, you must first have a drawing. Beginning a new drawing is just like creating any other file in Inventor; click a template and get to work.

1. On the Quick Access toolbar, click the New icon.

2. Make sure that 2013 Essentials.ipj is listed as the active project file near the bottom of the dialog box.

3. In the New File dialog box, select the Metric Template set on the left (see Figure 2.1).

4. Locate and double-click the ANSI (mm).idw template file.

This opens a new drawing page in the Graphics window and updates the Ribbon to present the Place Views tab and its tools.

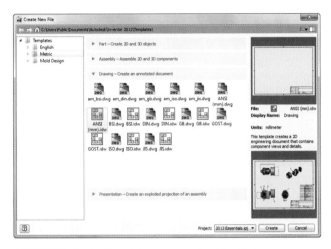

FIGURE 2.1 Templates are organized by file type.

Placing the Base and Projected Views

Now, you can begin the process of creating your drawing. For this exercise, you will use the marking menu tools.

Certification
Objective

1. Right-click in the Graphics window, and move up to select the Base View tool; or, right-click and drag up quickly to activate the tool.

2. In the Drawing View dialog box, click the icon to select an existing file at the end of the File drop-down.

3. Using the same Open dialog box you used in Chapter 1, "Connecting to Inventor's Interface," use the parts shortcut in the upper left; then open the Parts\Chapter 2 folder, and select the c02-01.ipt file.

   If you move your cursor over the drawing page, a preview appears, showing how the component will be sized and positioned on the page.

4. In the upper right of the Drawing View dialog box, click a few different options in the Orientation field and see the effect each one has on the preview. Make sure you end up back at the Front view.

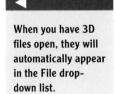

When you have 3D files open, they will automatically appear in the File drop-down list.

5. In the lower left of the dialog box, set the view scale to .7 and see the effect on the preview.

   At the bottom of many dialog boxes is an arrow icon. Clicking the arrow causes the dialog box to roll up to just its title bar. If a dialog box is in your way, use this option to make it easier to see your screen. Moving your cursor near the dialog box expands it to its full size, and reselecting the arrow holds the dialog box open.

6. Position the view similar to Figure 2.2, and click the mouse to place the drawing view.

7. Move your cursor down and to the right and to the upper-right corner, and click the mouse to place the views that are shown in Figure 2.3. You must right-click and select Create from the context menu to generate the views.

The ability to quickly place the base view and the projected views in one step is a great way to rapidly see how the drawing will lay out.

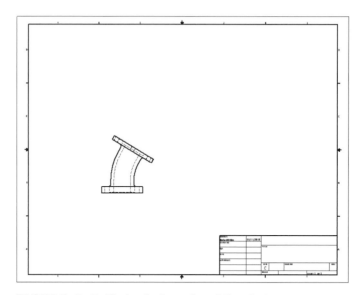

FIGURE 2.2 Placing the base view of the adapter

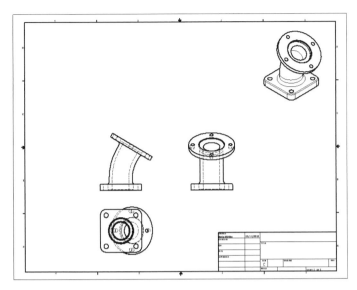

FIGURE 2.3 Locating the bottom, side, and isometric views

Placing a Section View

Section views are important for detailing machine parts and assemblies. Now, you will use the drawing views you have already created to define a section view using the steps in this exercise;

1. Make certain that the 2013 Essentials project file is active; then open c02-02.idw from the Drawings\Chapter 2 folder.

2. Start the Section tool from the Create panel on the Place Views tab.

3. Click the projected view to the right of the base view to choose it as the parent of the section view.

4. As you move over and around the parent view, you will see dotted lines that show the alignment to geometry in the parent view. When an inference line appears from the center of the top face (see Figure 2.4), click to start creating a section line.

5. Drag the line down through the part to define the section line, and click a second point to start creating the section view. See Figure 2.5. Right-click and select Continue from the context menu to finish defining the section line.

6. When the Section View dialog box opens, you can change the scale and view identifier. Move your mouse to the right to see a preview of the view. See Figure 2.6.

7. Click the location for the view to generate the view. Figure 2.7 shows the result.

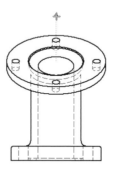

FIGURE 2.4 Inference lines will allow the section to be connected to geometry in the parent view

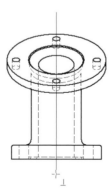

FIGURE 2.5 Include some overlap of the view in the section line length.

It is possible to define a section view of nearly any existing drawing view. Auxiliary views are very similar to section views but are defined by model geometry.

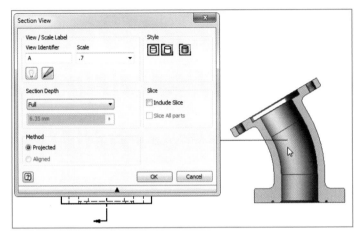

FIGURE 2.6 Placing the section view

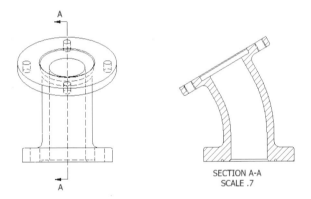

SECTION A-A
SCALE .7

FIGURE 2.7 The finished section view

Creating an Auxiliary View

Auxiliary views can be a challenge to define accurately with a 2D CAD system. In this case, you want to more easily detail the round face of the part, which does not align with a standard orthographic projection.

1. Make certain that the 2013 Essentials project file is active, and then open c02-03.idw from the Drawings\Chapter 2 folder.

2. Move your cursor near the base view, and when it highlights, right-click to bring up the marking menu and context menu with other tools.

3. Select Create View ➤ Auxiliary View from the context menu.

4. After the Auxiliary View dialog box appears, click the angled edge on the top of the base view, as shown in Figure 2.8.

5. Click to place your view, as shown in Figure 2.9.

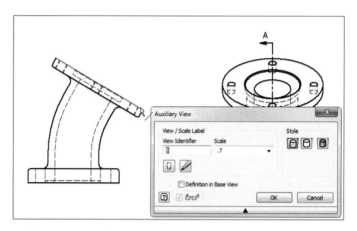

FIGURE 2.8 Selecting an edge for the projection reference

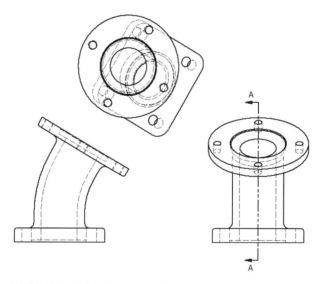

FIGURE 2.9 An easy auxiliary view

Creating a Detail View

The auxiliary view offers more clarity, but the section view offers an opportunity to see things otherwise obscured in an orthographic view. To do so, you'll create a detail view to add clarity:

1. Make certain that the 2013 Essentials project file is active, and then open c02-04.idw from the Drawings\Chapter 2 folder.

2. Start the Detail tool from the Ribbon or from the context menu.

3. Click the section view as the parent, which will open the Detail View dialog box.

4. Set Scale to 2 and Fence Shape to Rectangular; then, click near the O-ring groove, and drag the fence, as shown in Figure 2.10.

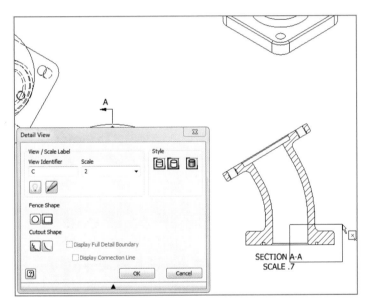

FIGURE 2.10 Setting up the focus of the detail view

5. Drag the new view to any open spot on the page, and click to place it.

Now that you have a collection of views, you should make the drawing look better as a whole. By moving, rotating, or even changing the appearance of the views, you can make a big difference.

Editing Views

The detail drawing is an important communication tool, so it is equally important to make it as clear as possible. In this section, you will focus on tools that will help you put the finishing touches on your drawings.

View Alignment

When views are created from other views, they inherit the appearance and scale of the parent view. Projected, auxiliary, and section views also inherit alignment from the parent view. Using this alignment and sometimes breaking the alignment make it simple to reorganize the drawing.

The simplest way to add clarity to a drawing is to shuffle the position of the drawing views. Doing this is a simple drag-and-drop operation.

You can even move drawing views to the opposite side from where they were created.

1. Make certain that the 2013 Essentials project file is active, and then open c02-05.idw from the Drawings\Chapter 2 folder.

2. Click and drag the base view up and to the left. In Figure 2.11, you can see how the views projected from it and their children keep the alignment of orthographic projection.

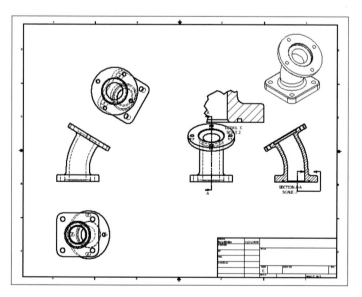

FIGURE 2.11 Moving a view does not break its proper alignment with others.

3. Move the right-side view and the section views to give them more space. Notice that they remain in alignment with the base view but do not change its position.

4. Move the detail view to the left of the title block, as shown in Figure 2.12.

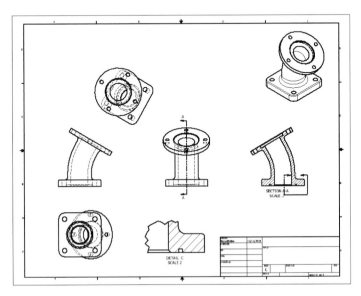

FIGURE 2.12 The detail view is free to be moved anywhere.

Moving views goes a long way toward making things look better. Sometimes, you might not want to maintain view alignment, or there may be times where it impedes the detailing of geometry.

Changing Alignment

It is possible to edit a view and break its alignment to a parent view using simple right-click options or using the tools in the Modify panel. You can also add alignment between views using tools from the same menus.

For example, to change the location of the auxiliary view, you need to break its alignment with the base view. To make it easier to dimension, you might want to rotate it as well. The Rotate tool will do both steps for you.

1. Make certain that the 2013 Essentials project file is active, and then open c02-06.idw from the Drawings\Chapter 2 folder.

2. Click the auxiliary view, right-click, and select Rotate from the context menu.

3. In the Rotate View dialog box, keep the rotation method By Edge, but set the alignment to Vertical and check that the direction is set to clockwise.

4. Click the straight edge on the right side of the view per Figure 2.13.

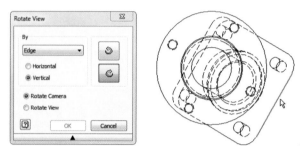

FIGURE 2.13 Rotating the drawing view can make detailing easier.

5. Click OK after the part has been rotated.

Now that the view has been rotated, you can detail it. But you might also want to reestablish an alignment with the base view to help others understand the geometry.

This exercise shows an uncommon and specialized drawing practice. It is built around the classic drafting idea of making it easier for people to recognize the geometry.

1. Make certain that the 2013 Essentials project file is active, and then open c02-07.idw from the Drawings\Chapter 2 folder.

2. In the Modify panel, expand the options under Break Alignment and select Vertical.

3. Click the rotated auxiliary view and then the base view. See Figure 2.14 for the results.

Now that you think you have enough room between the views and they are aligned, you will make some of them easier to interpret.

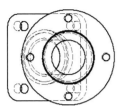

FIGURE 2.14 Views can be aligned to other views that were not originally their parent.

View Appearance

Changing how a view looks or is scaled can have a dramatic effect on how easy it is to understand.

Certification Objective

You can use a number of techniques, including removing hidden lines, adding shading, changing scale, or turning the visibility of selected entities off altogether. Any or a combination of these can be done without changing the accuracy of the view.

A quick way to add clarity is to remove hidden lines when they're not needed to understand the geometry of the component. In this exercise, you'll also add shading.

1. Make certain that the 2013 Essentials project file is active, and then open c02-08.idw from the Drawings\Chapter 2 folder.

2. Double-click in the auxiliary view (upper left) to open the Drawing View dialog box.

 3. Deselect the Style From Base check box.

 4. To the left of the Style From Base icon, click the Hidden Line Removed icon, and click OK to close the dialog box and apply the change.

5. Double-click the isometric view to edit it.

6. Click the Shaded icon.

7. Select the Display Options tab in the dialog box, and then remove the check mark from Tangent Edges.

8. Change the scale of the view to .5.

9. Click OK to update the drawing. See Figure 2.15.

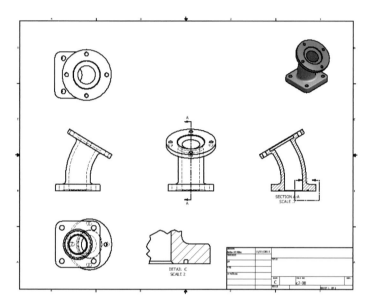

FIGURE 2.15 Updated drawing views

Removing unneeded hidden lines and adding shading can make the drawing easier to understand without adding more views. You could continue to make adjustments to the views for hours. Let's instead turn our attention to detailing the views you have.

Adding Detail to Drawing Views

You can add many things to a drawing to make it complete. Much of the work is tedious but necessary. Reducing the time it takes to add all of the detail is one of the unsung strengths of Inventor. In the next several pages, you will see some great examples of the productivity tools for detailing and dimensioning.

Most of the tools you will be using are located on the Annotate tab. This is one time where Inventor doesn't automatically change the active tab, because everyone uses the detailing tools differently.

Center Marks and Centerlines

You often need to add geometry to the drawing views to assist in locating dimensions. The next several exercises will walk you through the most commonly used tools for this purpose: center marks and centerlines.

Certification
Objective

Center marks make it easy to locate the center of a hole or radius.

1. Make certain that the 2013 Essentials project file is active, and then open c02-09.idw from the Drawings\Chapter 2 folder.

2. Click the Annotate tab in the Ribbon to change your active tab and see the Annotation tools.

3. Zoom in on the bottom view in the lower left of the drawing.

4. Click the Center Mark tool on the Symbols panel of the Annotate tab.

5. Move your cursor near the edge of one of the holes. When it highlights, click your mouse to place a center mark.

6. Repeat to place center marks on all four holes, as shown in Figure 2.16.

7. To connect the center marks, drag the center mark extensions toward each other.

Now that you've added and edited center marks for a basic rectangular hole pattern, you can move on to bolt circles.

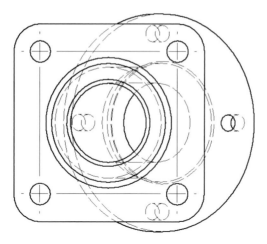

FIGURE 2.16 Placing center marks on holes
by simply picking them

Center marks and centerlines have many uses in Inventor. The ability to easily make and drag-edit the extents of these marks helps you save time cleaning up the drawing. They can also be used to show linear alignments between holes or even bolt circles.

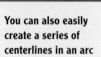

You can also easily create a series of centerlines in an arc based on a center.

1. Make certain that the 2013 Essentials project file is active, and then open c02-10.idw from the Drawings\Chapter 2 folder.

2. Zoom in on the rotated auxiliary view in the upper-left corner.

 3. Select the Centerline tool from the Symbols panel of the Annotation tab.

4. Begin by selecting the hole at 12 o'clock of the round face; then click the holes at 3 o'clock, 6 o'clock, and 9 o'clock, as shown in Figure 2.17.

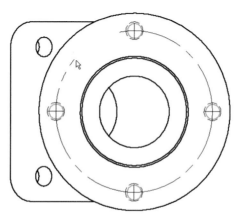

FIGURE 2.17 As you place the center marks
for the bolt circle, the centerline previews.

5. Finish the Bolt circle by clicking the 12 o'clock hole again; then right-click and select Create from the context menu. See Figure 2.18.

6. Press the Esc key to leave the Centerline tool.

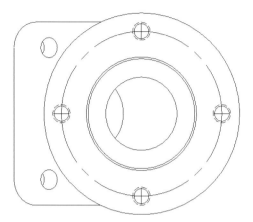

FIGURE 2.18 Generating a bolt circle with the Centerline tool

You can use the Centerline Bisector tool on straight or curved segments to find the midline between them. Straight segments do not need to be parallel.

1. Make certain that the 2013 Essentials project file is active, and then open c02-11.idw from the Drawings\Chapter 2 folder.

2. Zoom in on the section view on the right side of the drawing.

3. Select the Centerline Bisector tool on the Symbols panel of the Annotation tab.

4. Click the Orange lines in the lower left and lower right of the open portion of the part to place a straight segment.

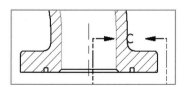

5. Click the green arcs on the left and right to place a curved segment.

6. Place another straight segment by picking the blue angled line at the top left and top right.
   You can also extend the straight segments beyond their original size for clarity. See Figure 2.19.

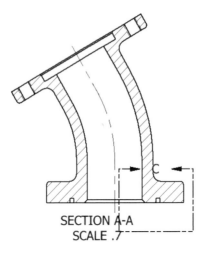

FIGURE 2.19 Placing a complex centerline bisector

While you are looking at the section view, you should improve its appearance.

Editing a Detail View Placement and Callout

Previously, you moved views around to improve the layout of the drawing. Callouts, section lines, and detail boundaries can also be edited to improve their clarity.

1. Make certain that the 2013 Essentials project file is active, and then open c02-12.idw from the Drawings\Chapter 2 folder.

2. Zoom in on the section view.

3. Select the letter *C* in the detail view boundary, and drag it to where the upper-right corner is now.
   In addition to relocating the callout, you can change the letter by editing the text.

4. Click the boundary. When the corners and center highlight, drag the center upward and to the right a short distance.

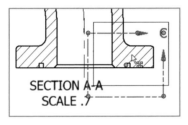

5. Select the *SECTION A-A* text, and drag it below the view so it is legible. Figure 2.20 shows the result.

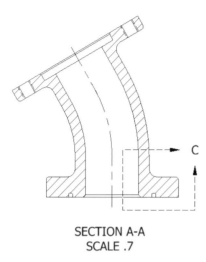

SECTION A-A
SCALE .7

FIGURE 2.20 Relocate the section view label for a better drawing view.

Now that the views have been updated with new or repositioned elements, you can turn your attention to dimensioning.

Dimensioning

If you've used traditional 2D CAD tools in the past for drafting, then the tools Inventor uses to apply dimensions should seem familiar but different at the same time. Rather than offer the user a broad array of dimensioning tools, Inventor uses intelligence that changes the type of dimension based on the selected geometry. Add the ability to change many options while you are placing the dimension, and you are sure to appreciate the power available to you.

The General Dimension Tool

To place individual dimensions in the drawing, the General Dimension tool will give you most anything you will need. For horizontal, vertical, aligned, radial, and diameter dimensions, you can just use this tool.

Certification
Objective

1. Make certain that the 2013 Essentials project file is active, and then open c02-13.idw from the Drawings\Chapter 2 folder.

2. Zoom in on the bottom view in the lower left of the drawing.

3. From the Dimension panel of the Annotate tab, select the Dimension tool.

4. Click the center marks in the lower- and upper-right corners (Figure 2.21).

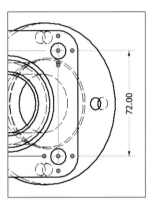

FIGURE 2.21 Click the center marks or other geometry to place dimensions.

5. Move your cursor to the right. You will see the dimension change to dotted lines when you have good spacing away from the geometry.

6. Click to place the dimension.

7. When the Edit Dimension dialog box appears, deselect the Edit Dimension When Created check box, and click OK to close the dialog box.

8. The General Dimension tool is still running. Click the radius in the lower-right corner, and notice that you are now creating a radial dimension. This dimension will snap at common angles.

9. Click to place the new dimension (Figure 2.22).

10. Pan up to the base view.

11. Click the upper edge of the base of the part and then the lower edge of the angle face. Notice that the dimension now turns to an angular dimension.

12. Place the new dimension to the side, as shown in Figure 2.23.

13. Select the top edge of the angle portion.

14. Place the linear diameter above the drawing view (Figure 2.24).

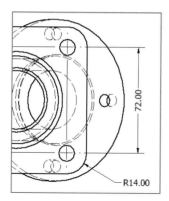

FIGURE 2.22 Adding a radial dimension

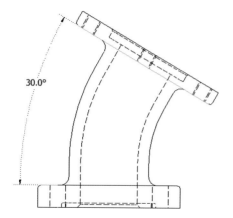

FIGURE 2.23 Even angular dimensions can be placed with the General Dimension tool.

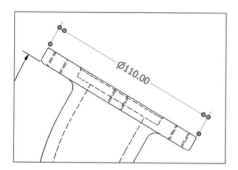

FIGURE 2.24 The geometry selected determines the type of dimension placed.

As you see, the General Dimension tool is incredibly flexible.

The Baseline and Baseline Set Dimension Tools

When you need to place several dimensions from the same datum, you can use the baseline dimensioning tools, Baseline and Baseline Set. The difference between the two is that Baseline Set keeps the placed dimensions' styles and precision linked along with other variables.

1. Make certain that the 2013 Essentials project file is active, and then open c02-14.idw from the Drawings\Chapter 2 folder.

2. Zoom in on the bottom view in the lower left of the drawing.

3. From the Dimension panel of the Annotate tab, expand the Baseline tool, and select Baseline Set.

   The tool starts immediately and waits for you to select the geometry.

4. Click the left edge of the square face, the top-left hole, the top-right hole, and the right edge.

5. Right-click and select Continue from the context menu to stop selecting geometry and begin placing dimensions.

6. When the preview of the dimensions appears, move them above the view.

7. Click to place the dimensions, and then right-click and select Create from the context menu to finish. See Figure 2.25.

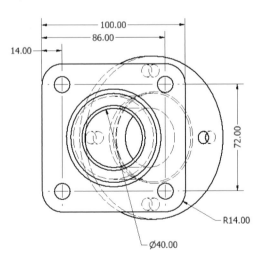

FIGURE 2.25 Placing multiple dimensions with the Baseline Dimension tool

The Chain and Chain Set Dimension Tools

Used commonly in architecture and when part cost is more important than high precision, the Chain dimensioning tools work very much like the Baseline tools but place dimensions in a series rather than from a common datum.

1. Make certain that the 2013 Essentials project file is active, and then open c02-15.idw from the Drawings\Chapter 2 folder.

2. Zoom in on the bottom view in the lower left of the drawing.

3. From the Dimension panel of the Annotate tab, select the Chain Dimension tool.
    The tool starts immediately and waits for you to select the geometry.

4. Click the top edge of the square face, the top-left hole, the bottom-left hole, and the bottom edge.

5. Right-click and select Continue from the context menu to stop selecting geometry and begin placing dimensions.

6. The preview of the dimensions will appear. Move them to the left of the view.

7. Click to place the dimensions; then right-click and select Create from the context menu to finish. See Figure 2.26.

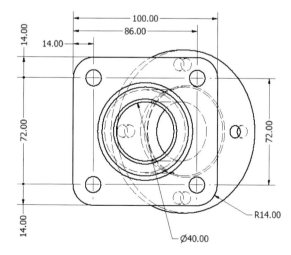

FIGURE 2.26 Chain dimensioning is much easier with a specific tool.

The Ordinate and Ordinate Set Dimension Tools

The Ordinate dimensioning tools work much like the Baseline tools but place the dimension value at the end of the extension line, as shown in Figure 2.27.

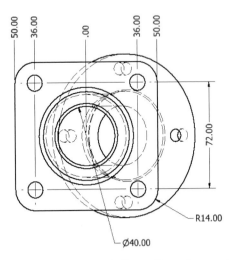

FIGURE 2.27 Ordinate dimensions work with the same workflow as chain dimensions.

The difference between the two is that the Ordinate Dimension tool prompts for the placement of a datum point in the view. The Ordinate Set dimension tool does not, and it therefore has a workflow that is essentially the same as that of the Baseline Dimension tool.

Editing Dimensions

When you placed the baseline dimension set, it was placed almost as a unit. The dimensions share many properties. You can still change these properties, and if need be, you can make a dimension independent.

1. Make certain that the 2013 Essentials project file is active, and then open c02-16.idw from the Drawings\Chapter 2 folder.

2. Zoom in on the bottom view in the lower left of the drawing.

3. Click the 72.00 dimension on the left.
   Once you've selected the dimension, two drop-downs will become active in the Format panel on the Annotate tab.

4. On the lower drop-down, change the Dimension style to Default—mm [in] (ANSI).

5. Right-click the 14.00 dimension in the baseline set, and select Detach Member from the context menu.

6. Double-click the 14.00 dimension.

7. When the Edit Dimension dialog box opens, select the Precision And Tolerance tab.

8. In the Precision field on the right, use the Primary Unit drop-down, and change the value to 1.1. Click OK. See Figure 2.28.

You can also delete a member and use the Arrange tool to bring the dimensions back in order.

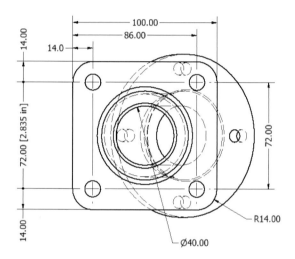

FIGURE 2.28 The dimensions can still be edited after being placed.

The Edit Dimension dialog box offers a lot of options that I didn't cover in the exercise, including tolerancing, fits, and even the ability to define an inspection dimension.

The Hole and Thread Notes Tool

Inventor's part modeling tools give you the ability to define holes by thread and their clearances for fasteners based on standards. To document these features, use the Hole and Thread note tool, which extracts the information directly from the feature.

Certification Objective

1. Make certain that the 2013 Essentials project file is active, and then open c02-17.idw from the Drawings\Chapter 2 folder.

2. Zoom in on the rotated auxiliary view in the upper left of the drawing.

3. Find and start the Hole and Thread tool in the Feature Notes panel of the Annotation tab.

4. Click the hole at 3 o'clock, and place the dimension above and to the right of the hole.

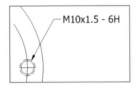

5. Press the Esc key to end the tool.

Now, you can edit the annotation to add detail.

6. Double-click the Hole note.

7. In the Edit Hole Note dialog box that appears (Figure 2.29), press the Home key or click the cursor before the text *<THDCD>*.

8. Click the Quantity Note icon to add the ability to display the number of like holes in the hole note. Then click OK. See Figure 2.30.

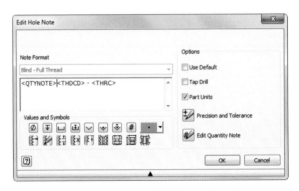

FIGURE 2.29 The Edit Hole Note dialog box

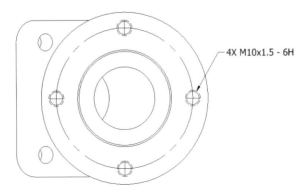

FIGURE 2.30 The Hole note can also display the number of holes.

Retrieving Model Dimensions

The parametric dimensions that are used in sketches to construct the parts can also be used to detail the part.

1. Make certain that the 2013 Essentials project file is active, and then open c02-18.idw from the Drawings\Chapter 2 folder.

2. Zoom in on the rotated auxiliary view in the upper left of the drawing.

3. Select the Customize tool from the Options panel on the Tools tab.

4. Select Full Menu from the Overflow Menu options drop-down in the lower left of the Marking Menu tab in the Customize dialog box, and then click Close.

    This will give you an expanded set of tools to use.

5. Right-click in the drawing view, and select Retrieve Dimensions from the context menu; or, select the Retrieve tool from the Dimension panel of the Annotate tab.

6. Once the Retrieve Dimensions dialog box is open, select one of the tapped holes in the drawing view.

    The drill diameter of the holes and the diameter of the bolt circle will be displayed. When Inventor needs more information from the user, an icon will appear in the dialog box with a red arrow. Often, the button will be depressed, and selection can be made immediately. Sometimes, you will need to select the button to begin picking entities.

7. Click the icon to select dimensions, and click the 90 mm diameter of the bolt circle.

8. Click OK to close the dialog box and include the dimension in the drawing view, as shown in Figure 2.31.

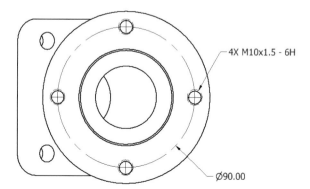

F I G U R E   2 . 3 1   Reusing model dimensions in the drawing view

TIP You might want to restore your Overflow Menu setting to the default. The short menu should provide you with the tools that you would normally use.

Associativity

The ability to create drawing views and annotate them should be clear by this point. But I haven't demonstrated the advantage of creating 2D parts from 3D geometry. When you edit a 3D part, any change is reflected in the 2D drawing.

This capability is key for productivity and demonstrates the final value of creating 2D drawings from 3D parts. The technique used in this exercise is not the most common way of making the changes, but it does show a technique that is available.

1. Make certain that the 2013 Essentials project file is active, and then open c02-19.idw from the Drawings\Chapter 2 folder.

2. Zoom in on the rotated auxiliary view in the upper left of the drawing.

3. Right-click the 90 mm diameter dimension. Because it is a model dimension, it will offer a special option.

4. Select Edit Model Dimension from the context menu.

5. In the Edit Dimension dialog box, change the value to 85 mm, and click the green check mark to make the change. See Figure 2.32 for the results.

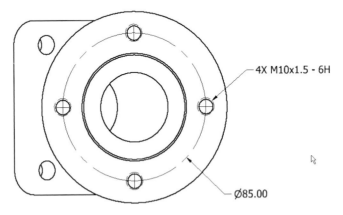

4X M10x1.5 - 6H

Ø85.00

FIGURE 2.32 Any change to the geometry of the model is reflected in the drawing views.

If you need more room for drawing views, you can change the size of the sheet by right-clicking the sheet callout in the browser.

The drawing view that you were focused on is not the only view that was changed. All views of this part on all pages in any drawing file will be updated based on this change.

Replace Model Reference

Many parts are variations on an existing design. Oftentimes, you will create a whole new part number for a minor difference. Inventor can copy an existing drawing and replace the source of its drawing views. In the following exercise, you will use the finished drawing for the c02-01.ipt file as the source for a new drawing based on a new part.

1. Make certain that the 2013 Essentials project file is active, and then open c02-20.idw from the Drawings\Chapter 2 folder.

2. Start the Replace Model Reference tool from the Modify panel on the Manage tab.

3. In the Replace Model Reference dialog box, select the current drawing, and then click the Browse icon. In this case, there is only one model referenced.

4. Browse, select the Parts\Chapter 2\c02-03.ipt file, and click Open.

5. When the warning that you are changing the model appears, click Yes; then click OK to update the drawing and see the change, as shown in Figure 2.33.

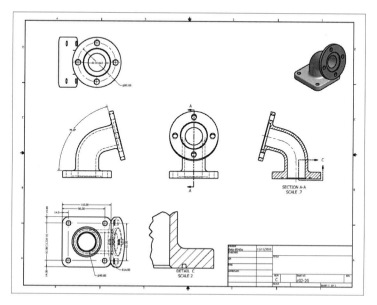

F I G U R E   2 . 3 3 Some dimension cleanup is necessary, but this was finished in seconds.

Some annotation might be lost or have to be relocated when making a change like this. For parts that have the same geometry but different sizes, this technique can offer a huge time-savings for detailing the new part.

THE ESSENTIALS AND BEYOND

There are countless ways to approach creating your detail drawings, and Inventor maintains Autodesk's legacy of flexibility but adds a level of simplicity that's not possible when creating 2D drawings from scratch.

ADDITIONAL EXERCISES

▶ Experiment with all the tools on the Place Views tab.

▶ Use DWG templates to create drawings that can be read using AutoCAD.

▶ Work with individual dimensions and dimension sets to see what method you might typically want to use.

▶ Try breaking the alignment of the section view to see that it still maintains its definition.

CHAPTER 3

Learning the Essentials of Part Modeling

Part modeling *is typically* the primary focus when beginning to learn Autodesk Inventor 2013. The tools involved are the ones you will spend a majority of your time using, but they are pretty straightforward.

In this chapter, you will look at the most commonly used tools and understand how to construct one of the most challenging things you can do with Inventor, the parametric sketch. It is challenging because, unlike most features, there aren't many limitations. You will have to use your judgment about what to include and exclude from the sketch. As you gain experience with your designs, this will become second nature.

Here is the process of beginning a new part, in shorthand: create a sketch that defines the most important elements of the part's geometry, and then turn that sketch into a 3D shape. Along the way, you'll add features that complete the part. That really is how simple the process is.

▶ **Defining a parametric sketch**

▶ **Creating 3D geometry: the parametric solid model**

Defining a Parametric Sketch

The term *parametric* refers to the process of using values (parameters of the object being designed) to define the size, location, and other properties of the object's features. In the sketch, this is most easily recognized in the dimensions that you add, but other types of constraints add control to the geometry.

In this chapter, you will create the part shown in Figure 3.1. Let's begin by constructing your first sketch. You will control the sketch by placing dimensional constraints (size values) as well as geometric constraints (vertical, parallel, and so on).

FIGURE 3.1 The 30-degree adapter you will build

Creating a Sketch

Certification
Objective

When Inventor begins a new part file, it starts in the sketching environment by default.

1. Open the New File dialog box from the Application menu, the Quick Access toolbar, or the Launch panel of the Get Started tab.

2. Select the Metric folder under Templates, and double-click the Standard (mm).ipt template.

   Inventor doesn't assume you will sketch or on what plane you would like to begin sketching.

3. Click the Create 2D Sketch tool in the Sketch panel.

   Three planes appear in the center of the design window. You must select one of these sketch planes.

4. For this exercise, select the plane that is parallel with the x- and y-axes, as shown in Figure 3.2

   After the plane is selected, Inventor aligns your view to look directly at the sketch plane.

4. Start the Line tool. Right-click in the Graphics window, and select the Line tool from the marking menu or find it in the Draw panel.

5. Begin the line segment by clicking below and to the left of the projected center point (the green dot above the line segment shown in Figure 3.3) of the part.

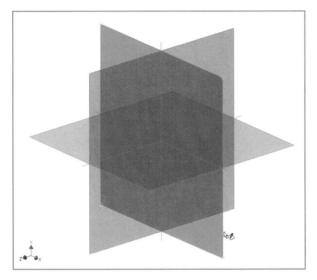

FIGURE 3.2 Selecting the plane for sketching

For clarity, the Grid, Axes, and Minor grid lines and Axes options in the Application Options dialog box's Sketch tab are disabled.

6. Begin moving your cursor to the right. You will see a dimension for the length of the line and the angles appear as you move around.

7. Position your cursor to the right of the part center with the line horizontal.

   As you are sketching, icons will appear that look like the constraint icons in the Constrain panel. In this case, the Horizontal constraint icon will appear when the line is horizontal.

8. The dimension value is highlighted, so type 60 to specify the length value for the line (Figure 3.3).

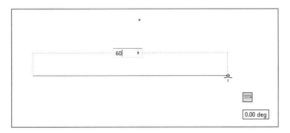

FIGURE 3.3 Entering a dimension value will lock in that size.

9. Press the Tab key to set that length value. When your line is horizontal, click the mouse to create the segment. Note that the Line tool is ready to create another segment.

10. Enter a value of 30 for the next segment, and use the Tab key to set the angle to 100. You might have to pan or zoom to see the second segment.

11. Click in the Design window to place the new segment, and notice the new dimensions that appear, as shown in Figure 3.4.

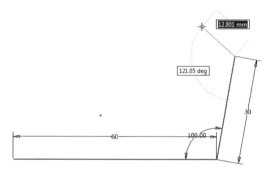

FIGURE 3.4 Adding values to dimensions creates the parametric dimensions on the fly.

12. Move your cursor to the left. Look for an icon that shows the new segment is parallel to the first. Move to the left until you also see a dotted "inference" line that aligns you with the starting point of the first segment.

13. Click the mouse when your line is parallel with an aligned endpoint, as shown in Figure 3.5.

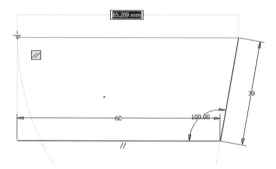

FIGURE 3.5 If you don't specify a dimension value, no dimension is added to the sketch.

14. Now move your cursor down to the original start point. The point will highlight, and a glyph will appear, showing that there will be a connection called *coincidence* between the points, which means the

endpoints of the line segments are attached to one another and will move together.

15. Click the mouse to connect the lines. See Figure 3.6.

16. Press the Esc key to finish the Line tool.

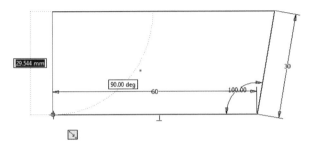

FIGURE 3.6 The initial shape is done.

Now that you have your initial shape, you can modify it to be the correct shape and size. Getting your sketch close to correct is good, but as long as you have the elements you need, you can modify it to be correct.

Adding and Editing Geometric Constraints

Creating the sketch added sizes and created relationships between sketch elements. At this point, you could easily turn this into a solid model.

Certification Objective

What you really need, though, is a square centered on the center point of the part. Now, you will make modifications to fix this and see some other options along the way:

1. Continue using the drawing from the previous exercise or make certain that the 2013 Essentials project file is active, and then open c03–01.ipt from the Parts\Chapter 3 folder and double-click Sketch1 in the browser to make it active.

2. On the status bar, click the Show All constraints icon, or simply press the F8 key. Figure 3.7 shows the result.

This will display the relationships between the elements. The yellow blocks at the corners represent coincidence.

> When the sketch (or elements of it) are selected in the Graphics window, you can access constraint visibility through the right-click context menu.

3. Hover over the Perpendicularity icon in the lower-left corner, and it will highlight the segments that are constrained.

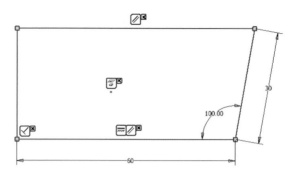

FIGURE 3.7 You can display the geometric and dimensional constraints of the sketch.

4. Hover over the other icons to see what constraints exist beyond the dimension values. Be sure to include the icons in the corners, which will expand.

 5. Press F9 or select Hide All Constraints on the status bar to turn off the constraints display.

6. Select the angular dimension.

7. Once it is highlighted, press the Delete key to remove it.

 8. On the status bar, select the Show All Degrees Of Freedom icon to see the display shown in Figure 3.8.

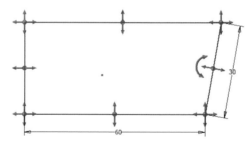

FIGURE 3.8 Showing the degrees of freedom of the sketch can help you understand where to add constraints.

This will display bold, red arrows showing how elements of the sketch can still be moved with a simple click and drag. As you apply constraints, these degrees of freedom will disappear. On the right end of the status bar, you can see how many constraints need to be added to have a fully constrained part.

9. Click the upper-right corner of the sketch, and drag it to the left. Notice how the length still remains 30 mm. You can move other elements as well.

10. In the Constrain panel, find the Perpendicular Constraint tool and start it.

11. Click the top and right lines to apply the constraint.
    The curved arrow will disappear, showing that the angle of the right line can no longer change.

12. Find and start the Horizontal constraint.

13. Click the center point of the sketch, and then move to the middle of one of the vertical lines.

14. When its midpoint highlights (turns to a green dot), click it, and the two points will align to one another.
    You will also see a dramatic effect on the degrees of freedom of the sketch. A more subtle change will be a difference in the color of the lines that are now constrained.

15. Find and start the Vertical constraint.

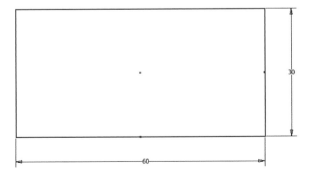

16. Select the center point and the midpoint of one of the horizontal lines.
    All the degrees of freedom of the sketch will disappear, and the status bar will report that the sketch is fully constrained, as shown in Figure 3.9. The icon for the sketch in the browser will also change, adding a tack when it is fully constrained.

FIGURE 3.9 The fully constrained sketch

> In cases where you want a square centered in the sketch, you might consider using the Polygon tool with four sides or a Two Point Center Rectangle.

17. Press the Esc key to end the Vertical constraint tool.

18. You can use the Hide All Degrees Of Freedom tool in the status bar to turn off the display for future sketches, if you would rather not see them.

When you're creating sketches for your own geometry, you won't create them as loosely as this exercise. It is also important that you create your sketches to be the approximate size of the finished shape. This will save you additional editing when you begin editing the dimensions. If you create a sketch proportional to itself but accidentally make it the wrong size, it will automatically scale itself based on the first parametric dimension you add to the sketch.

Editing Dimensions

Certification
Objective

This is the step where you will begin to see the value of parametrics. Creating things to be the right size today doesn't mean you won't need to change that size tomorrow. That is why approximating the geometry and adding precise dimensions that change it is a great workflow.

1. Double-click the 30 dimension to edit its value.

2. When the Edit Dimension dialog box appears, change the value to 40, and click the green check mark to make the change.

3. Edit the same dimension, and instead of entering a value, click the 60 dimension to link the two together.

4. Click the check mark to accept the new value.
   Note that the value has been set to d0, and an *fx* now appears in front of its value in the sketch. The d0 value refers to the name of the dimension it is linked to, and fx denotes that this dimension is based on another parameter.

5. Double-click the bottom dimension to edit it.

6. Enter the value of **Width=100**, and accept it.
   This changes the name of that dimension from d0 to Width and updates the sketch to show the new size. You might have to zoom out to see the sketch now.

7. Double-click the vertical dimension to see that it now refers to a value of Width; it must remain associated to the other dimension regardless of what changed.

8. Close the Edit Dimension dialog box without changing anything.

9. Select the vertical dimension again, and delete it.

TIP You can place dimensions in different ways. By default, dimensioning displays the value and any applied tolerances. Here are some other options:

⊢ˣ⊣ Value	100	
⊢ᵈ⊣ Name	Width	
⊢ᵈ⁼ˣ Expression	Width = 100 mm	
⊢⁺⁄₋ Tolerance	100 ⁺⁰·⁵₋₀.₂	
⊡ Precise Value	100.02503694	

10. Locate the Equal constraint (or press the = key), and select first a horizontal and then a vertical line to link their sizes.

11. After the sketch updates, use the marking menu to select Finish Sketch (or select it from the Exit panel on the Sketch tab).

Once the sketch is finished, the Ribbon updates to show the 3D Model tab, and the sketch moves to the Home view. If you've orbited the sketch while you were editing it, the view might not change orientation.

Creating a Pattern in a Sketch

Using a pattern to replicate sketch geometry is much faster than creating and dimensioning separate instances of the geometry.

Certification Objective

1. Make certain that the 2013 Essentials project file is active, and then open the c03-02.ipt file from the Parts\Chapter 3 folder.

2. Double-click the icon next to Sketch1 in the browser.

3. Click the Rectangular Pattern tool in the Pattern panel of the Sketch tab.

4. When the Rectangular Pattern dialog box opens, pick the circle in the sketch for the geometry that will be duplicated.

5. Pick the selection icon for Direction 1, and then pick the bottom horizontal line in the sketch to set the direction.

6. Set the count for Direction 1 to 5, and set Spacing to 20 mm.

7. Pick the selection icon for Direction 2, and pick the vertical line on the left.

8. Click the Flip icon to change the direction of the second pattern direction.

9. Set the count for Direction 2 to 5, and set Spacing to 20 mm, as shown in Figure 3.10.

10. Click OK to create the pattern.

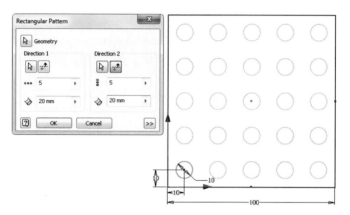

FIGURE 3.10 Building a pattern of holes

The two directions in the Rectangular pattern do not need to be perpendicular to each other. They can even be collinear going in opposite directions.

Using Sketches for Concept Layout

Using 2D to test a concept is not new. Using a combination of parametric dimensions and sketch constraints can bring this idea up-to-date and offer quick changes to test an idea.

1. Make certain that the 2013 Essentials project file is active, and then open the c03-03.ipt file from the Parts\Chapter 3 folder.

2. Double-click the Sketch icon in the browser to activate the Linkage sketch.
   Under the sketch in the browser, you will see three instances of a sketch block named Link 1 and two circles and a line that also reside in the sketch.

3. In the Layout panel of the Sketch tab, click the Create Block tool.
   Adding entities to a sketch block allows you to group the sketch elements together without creating a series of sketches.

4. Click the pink circles and line.

5. Pick the Insert Point select tool, and then pick the center of the right circle.

6. Enter Link 2 for the block name, and click OK to create a new sketch block.

7. Click the center point of the circle on the right of the new block, and drag it to the center point at the top of the vertical red bar on the right (see Figure 3.11); release when it shows it is creating a constraint between centers.

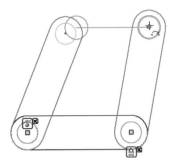

FIGURE 3.11 Dragging sketch constraints between sketch blocks

8. Click and drag the center of the new block on the left to the center point of the vertical bar on the left.

9. Click and drag one of those endpoints to see how the linkage moves.

10. Double-click one of the instances of Link 1 in the Design window.

11. When the sketch geometry appears, click the 50 mm dimension and change it to 80 mm.

12. Click the Finish Edit Block icon on the Exit panel of the Sketch tab.

13. Move the links using the new part length.

Finish Edit Block

Sketch blocks can be simple or complex as needed to help you predict how a mechanism will work. Sketch blocks are quick, and the sketch can be used to build 3D data and speed up the development of an assembly.

Understanding Sketch Alerts

Occasionally, a sketch may be missing constraints that will affect your ability to create geometry. Inventor includes tools that help you locate and correct these missing constraints.

Certification Objective

1. Make certain that the 2013 Essentials project file is active, and then open c03-04.ipt from the Parts\Chapter 3 folder.

2. Click the Extrude tool on the 3D Model tab in the Create panel.

A preview of a surface model will appear. Inventor will extrude a surface when there is no closed profile. Even though the dimensions create an intact perimeter, Inventor needs line and arc endpoints to be coincident to form a completely closed sketch.

Instead of stopping the command, you can force the issue by telling Inventor that you want to create a solid.

3. In the Extrude dialog box, click the Solid option in the Output group.

This will cause an error to be presented in the dialog box. It will also activate an icon at the bottom of the dialog box with a red cross. This is called the Sketch Doctor.

4. Click the Examine Profile Problems tool.

The message in the Sketch Doctor – Examine dialog box will tell you about the open loop in the profile.

5. Click Next to continue.

6. In the Sketch Doctor – Treat dialog box (Figure 3.12), click the Close Loop treatment option, and then click Finish.

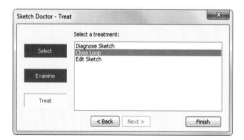

FIGURE 3.12 Selecting the treatment option for sketch problems

7. Click Yes to combine the points and make a closed sketch.

8. Click OK to close the dialog box informing you that the sketch is now closed.

9. Start the Dimension tool and pick the green line in the sketch.

A new Redundant Points dialog box (Figure 3.13) will appear, and in the Design window these points will be highlighted.

10. Pick a place to position the new dimension.

This will launch the dialog box shown in Figure 3.14, warning you that adding the dimension will over-constrain the sketch. This means that this new dimension is redundant. The dialog box gives you the option of just canceling the placement of the dimension or placing it as an associative dimension.

11. Click Accept to place the driven dimension in the sketch.

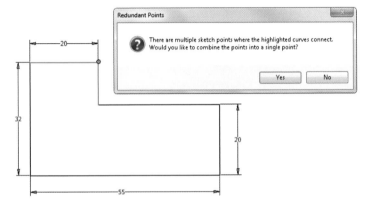

FIGURE 3.13 The problem points are highlighted in the Design window.

FIGURE 3.14 A dialog box warns you if you are placing dimensions that cannot be used in the sketch

The power of placing geometric and dimensional constraints can quickly turn against you if you're unable to overcome mistakes. The Sketch Doctor and warnings regarding an over-constrained state help you keep working by fixing the problems and moving on.

Now, it's time to start making some 3D geometry.

Creating 3D Geometry: The Parametric Solid Model

In Chapter 2, "Creating 2D Drawings from 3D Data," you created 2D drawings from an existing solid model. You also made a change to that model that updated all of the drawing views by accessing the value of a dimension. That dimension was part of a sketch that defined the placement of features also based on parameters.

These parameters can be as simple as a fillet radius, or they can be based on the clearance hole standard for a particular type of fastener. Either way, when you work with a parametric model, you open up a new world of control over your designs.

Now, you have a sketch that represents the geometry from which you'd like to make your first feature. When you start a part, the first feature you create will be a sketched feature, most commonly an extrusion or a revolved feature.

 N O T E   The following modeling exercises describe the procedures using the mini-toolbar and in-window tools. All the tools and options are also available in the Ribbon or in dialog boxes.

Extruding the Base Feature

Certification
Objective

The first feature in a part is referred to as the *base feature*. It is usually beneficial to define as much geometry as possible with the base feature.

1. Make certain that the 2013 Essentials project file is active, and then open c03-05.ipt from the Parts\Chapter 3 folder.

2. In the Design window, click one of the sketch lines. As you do so, commonly used tools will appear.

3. Click the Create Extrude tool on the left.

4. An initial preview of the extrusion will appear with a mini-toolbar appearing in the middle of the screen and a full dialog box. Both show the current value for the extrusion distance.

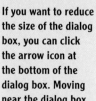

If you want to reduce the size of the dialog box, you can click the arrow icon at the bottom of the dialog box. Moving near the dialog box will expand it temporarily, or you can click the down arrow while the dialog box is expanded to lock it at full size again.

5. Click and drag the arrow at the end of the extrusion until the value in the mini-toolbar reads 15; or, highlight the value in the mini-toolbar, and enter 15. See Figure 3.15.

6. Moving near the mini-toolbar also displays additional options. Click the green check mark, which represents OK, to create the feature.

The extrusion is generated on the screen, and a feature named Extrusion1 appears in the browser.

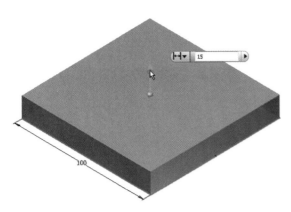

FIGURE 3.15 The option in the dialog box updates while you drag the direction arrow.

Setting the Material and Color

You can do things such as establish the material your part is made from to monitor its weight or inertial properties. You can even pick a specific color if you like.

1. In the browser, right-click the part icon at the top, and select iProperties from the context menu.

2. In the dialog box, go to the Physical tab, and use the Material drop-down to select Cast Iron as the material. Notice that it updates the mass; make note of what that value is.

3. Click OK to accept this change.

4. In the Quick Access toolbar, locate the drop-down that shows As Material for the part's color. Select a color you like, and feel free to change it whenever you want.

Setting the material from which the parts will be made from is very important for estimating costs, weight, and so on. Color changes can help highlight components or make them appear more realistic.

c3-02.ipt

◀

At various times while working in this and other chapters, you should go back to the Physical tab and update the properties to see how changes affect these values.

TIP Inventor also features a few different color libraries. If you intend to work with other Autodesk applications like 3DS Max and you want maintain consistency in the color, you should use a color from one of the Autodesk libraries.

Reusing Sketch Geometry

On a complex part where there are many features created from the same base plane, you can sometimes use a single sketch to define more than one feature.

1. Continue using the drawing from the previous exercise or make certain that the 2013 Essentials project file is active, and then open c03-06.ipt from the Parts\Chapter 3 folder.

2. Click the small + icon in the browser next to Extrusion1. This exposes the sketch that was consumed by that feature.

3. Right-click the sketch, and select Share Sketch from the context menu.
   The sketch now appears both under and above the extrusion feature. It also becomes visible in the Design window so other features can use it. In this case, you need to add to it first.

4. Click the top face of the extrusion. When you do, icons will appear for Edit Extrude, Edit Sketch, and Create Sketch.

5. Click Edit Sketch.

6. The Center Point Circle tool is on the marking menu and in the Draw panel. Start it, and draw a circle based on the part center point with a diameter of 40.

7. Draw a larger circle using the same center with no diameter dimension; then, press the Esc key to stop the Circle tool. See Figure 3.16.

8. Start the Dimension tool, and click the two circles. This will create a dimension showing the distance between the perimeters. Set it to 8, and finish the Dimension tool.

Your sketch might rotate so that you're looking directly at it. To create images in an isometric view, I have disabled this feature in the application options to show that sketches can be edited in this way, as well.

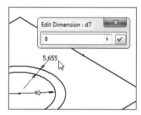

9. Finish editing the sketch.

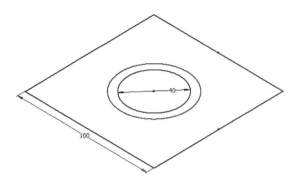

FIGURE 3.16 Adding new geometry to an existing sketch

Now, you can create more sketch geometry that is based on the center of the part. This is why it can be very useful to think about centering a part on the origin planes.

1. Continue using the drawing from the previous exercise or make certain that the 2013 Essentials project file is active, and then open c03–07.ipt from the Parts\Chapter 3 folder.

2. Expand the Origin folder in the browser.

3. Click the XY plane. This will highlight the plane in the Design window and present you with the Create Sketch icon.

4. Click the icon to start a new sketch on that plane.

5. The center point of the part will automatically project into the new sketch, but it will be difficult to see it from the front.

6. Right-click in the Design window, and select Slice Graphics (or press the F7 key) to remove some of the part to expose the new sketch. You can also pick the Slice Graphics icon in the status bar.

7. Start a new line at the part's center point, going straight up click roughly at 30 mm.

8. Move your mouse to the endpoint of the line you just created, click, and drag an arc tangent to the last line, as shown in Figure 3.17.

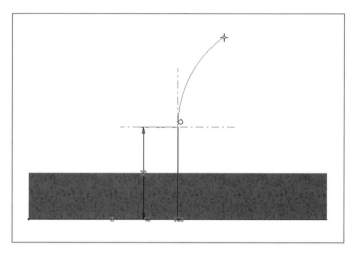

FIGURE 3.17 A tangent arc can be created using the Line tool.

9. Now add a straight line that is tangent to the arc. A glyph will appear that shows when the line is tangent. Press the Esc key to finish the Line tool.

10. Select Create Dimension from the marking menu.

11. Click the arc. When the radius dimension preview appears, right-click, and from the Dimension Type menu select Arc Length. Place the dimension and set the value to 50.

12. Click the second straight line, right-click, and click Aligned. Place the dimension and set the length of the line to 30.

13. Now click the first line and then the second line. Because they are not parallel, an angular dimension will appear. Position it as shown in Figure 3.18, and set its value to 30.

14. Finish the sketch.

Getting feedback on constraint conditions like tangency during the sketching process helps you know you're getting the control you need from the beginning.

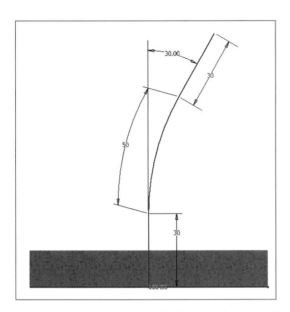

FIGURE 3.18 Refining the sketch with parametric dimensions

Connecting to Other Sketch Data

Linking to geometry in other sketches helps keep things in line.

Certification Objective

1. Continue using the drawing from the previous exercise or make certain that the 2013 Essentials project file is active, and then open c03-08.ipt from the Parts\Chapter 3 folder.

2. Start a new sketch on the XY plane.

3. Use the Project Geometry tool on the marking menu or in the Draw panel, and click the top 30 mm line to link it into the current sketch.

4. In the Draw panel, expand the Rectangle tool, and start the Three Point Rectangle tool.

5. Start the rectangle at the top end of the projected line. Drag to the right so it is perpendicular to the projected line, and click.
   To properly locate this sketch, you should create it so it is below the end of the 30 mm line. This will make placing the diameter more difficult, so temporarily create it in a different location.

6. Place the third point above the rest of the sketch at approximately 10 mm, as shown in Figure 3.19.

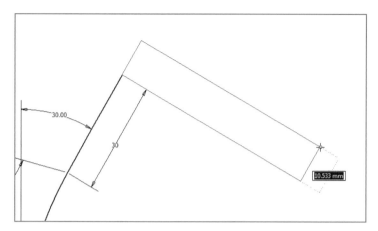

FIGURE 3.19 A three-point rectangle can be placed at any initial angle.

7. Stop the Rectangle tool by pressing the Esc key or by selecting
   Cancel from the marking menu.

   The new rectangle will eventually be revolved so that even though
   you are looking only at the side, it would be best to dimension it as
   its final diameter. Using the Centerline override, you can get the
   dimension tool to recognize this.

 8. Click the short edge of the rectangle that is collinear with the 30 mm
   line, and then pick the Centerline override from the Format panel.

9. Start the Dimension tool, and click first the centerline and then the
   opposite end of the rectangle.

10. Place the dimension, and set its value to 110 mm. See Figure 3.20.

11. End the Dimension, tool and drag the top edge of the rectangle down
    so it is positioned below the end of the 30 mm line.

12. Restart the Dimension tool, and place a 10 mm dimension for
    the thickness by picking long lines at the top and bottom of the
    rectangle.

13. Finish the Dimension tool and editing the sketch.

Let's keep building on your part. Moving forward, you will be switching files
because some of the basics will be prepared for you so you can focus on new things.

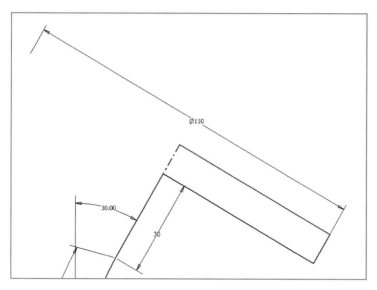

FIGURE 3.20 A linear diameter can be placed in the sketch.

Creating a Revolve Feature

For many users, revolving is the second most common thing to create with a sketched feature after extruding. If you make turned components, it may be your number-one tool, although I think you'll like the Shaft Generator covered in Chapter 8, "Advanced Assembly and Engineering Tools."

Certification
Objective

1. Make certain that the 2013 Essentials project file is active, and then open c03-09.ipt from the Parts\Chapter 3 folder.

2. Click an element of the rectangle in Sketch3, and choose the Create Revolve icon.

3. Because there are multiple visible sketches with more than one complete profile, you will need to click the shape of the rectangle.

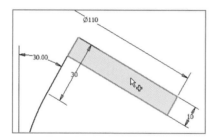

4. Now, click the Axis icon on the mini-toolbar to let Inventor know you're done selecting profiles.

5. Select the centerline that is in line with the 30 mm dimension.
   This will immediately produce a preview of the new feature (Figure 3.21). Note that an arrow allowing for the changing of the angle is available.

6. Finish the Revolve tool by clicking the green check mark.

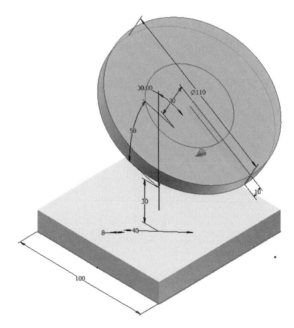

FIGURE 3.21 A preview of the Revolve tool

Now you can complete the main features of the part.

Creating Sweep Features

In this exercise, you'll create two features. The first will establish the outer portion and will be hollow. The second will cut through the top and bottom plates.

1. Make certain that the 2013 Essentials project file is active, and then open c03-10.ipt from the Parts\Chapter 3 folder.

2. Start the Sweep tool from the Create panel on the 3D Model tab.
   Adding the circles to Sketch1 has made it a multiloop sketch, so there is more than one option; the button for profile selection in the dialog box will have a red arrow. This means Inventor needs you to make a selection.

3. Click in the space between the concentric circles.

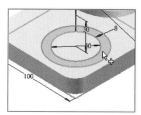

4. Because Optimize For Single Selection is selected, Inventor now wants you to click the path. Select the part of the lines and arc going to the disk on top, as shown in Figure 3.22.

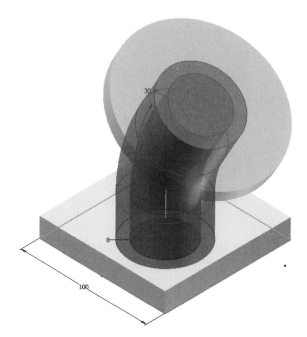

FIGURE 3.22 A preview of the first Sweep tool

5. Click OK in the dialog box to finish the tool.
   This consumed the sketch that you used for the sweep path.

6. Expand the Sweep feature in the browser.

7. Right-click and make Sketch2 visible.

8. Start the Sweep tool, and use the inner circle for your profile.

9. Select the same path sketch.

10. Set your option in the dialog box to Cut.

11. Click OK to create the feature.

12. Turn off the visibility of the two sketches. Figure 3.23 shows the part at this point.

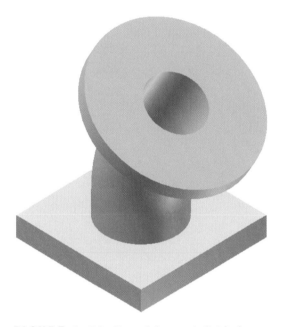

FIGURE 3.23 Most of the part is finished.

By sharing a sketch that defines more than one feature, you can control the relationships between the various sketches. Now, you will focus on adding a few placed features to complete the part.

Leveraging Primitives

So far, you've created simple shapes such as boxes and cylinders using sketched features, but that was more about sketching than creating solid models. Primitive shapes such as boxes, cylinders, and spheres are easier to place as intact features. You might even want to re-create the model using these primitives to explore that option. For now, you will just use one to add geometry to the model.

1. Make certain that the 2013 Essentials project file is active, and then open c03-11.ipt from the Parts\Chapter 3 folder.

2. Use the menu in the Primitives panel to start the Cylinder tool.

3. Select the top face of the revolved feature for a plane to place the cylinder on.

   This will automatically project the geometry of that face including the centerpoint of the concentric circles.

4. Pick the center of the circles for the center of the cylinder and drag the diameter.

5. Enter a diameter of **55**, and hit Enter.

   This automatically switches the view position of the model and starts the Extrude tool.

6. Drag the distance arrow down and back into the part to convert the operation from a Join to a Cut, as shown in Figure 3.24.

7. Set a depth of 5 mm, and click OK to place the new feature.

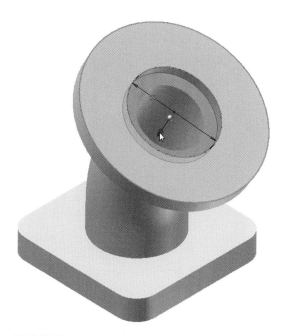

FIGURE 3.24 Primitives accelerate the process of creating basic shapes.

Looking at the browser, you won't see a special feature callout, and editing the extrusion that was created is no different from if you had gone through the process step by step. As with many of the most productive tools in Inventor, this is just an easier way to do things that allows you to stay focused on what you're creating, not how you're creating it.

Adding an Edge Fillet

One of the best practices for modeling is *not* to include incidental fillets in the sketch. It is better and easier to place fillet features in the model instead.

1. Continue using the drawing from the previous exercise or make certain that the 2013 Essentials project file is active, and then open c03-12.ipt from the Parts\Chapter 3 folder.

2. Click one of the short, vertical edges on the square base of the part.

3. When the icons appear for the most common options, select the Create Fillet tool.

4. Click the remaining three short edges.

5. An arrow icon will follow your selections. Drag it to create a 14 mm radius on the four corners, as shown in Figure 3.25.

6. Thanks to marking menus, a quick right-click and drag to the right will give you the same effect as clicking OK. Try this to complete the fillet tool.

It is possible to click faces and edges through a part by hovering over their location. You can also orbit while in a command without interrupting it.

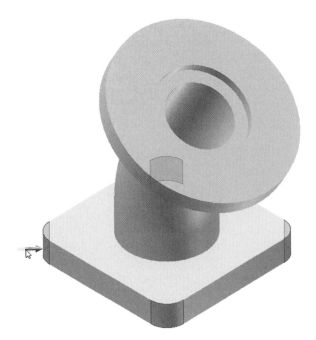

FIGURE 3.25 The fillets will develop as you drag them.

Fillets are very common, and in Chapter 7, "Advanced Part Modeling Features," I cover additional techniques for placing them. It is also possible to place fillets with several different values at the same time.

Adding Edge Fillets with Multiple Radii

Fillets are another critical feature in machine parts. Fillets have many options, but for this chapter, you will focus on the edge fillet and placing fillets of more than one radius using a single feature.

1. Make certain that the 2013 Essentials project file is active, and then open c03-13.ipt from the Parts\Chapter 3 folder.

2. Click the round edge where the sweep meets the extrude.
    This presents a choice of the Create Fillet or Create Chamfer tool.

3. Select the Create Fillet tool.

4. Add the edge where the sweep feature meets the revolved feature.

5. Drag or enter a radius value of 6.

6. In the Fillet dialog box, click where the words *Click to add* appear.

7. This adds a new row for using another radius size. Click the radius value of the new row, and change it from 6 to 2.

8. Click the outer edges of the extrusion and revolved features near the first set of fillets. See Figure 3.26.

9. Click OK to finish applying the fillets.

The option of placing a fillet or a chamfer on a sharp edge makes perfect sense, so it is great that you have access to the option by simply picking a sharp edge.

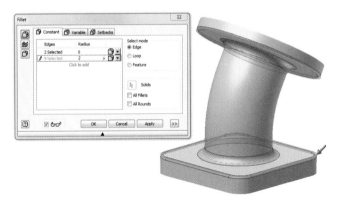

FIGURE 3.26 You can place fillets with different radii in one fillet feature.

Applying a Chamfer

Certification
Objective

Chamfers and fillets have similar locations, but their options are quite different. For this exercise, you will use two of the three options.

1. Make certain that the 2013 Essentials project file is active, and then open c03-14.ipt from the Parts\Chapter 3 folder.

2. Click the sharp edge on the interior of the revolved feature.

3. Select the Create Chamfer tool. The default is 45 degrees. Drag the Chamfer value to 2 mm, as shown in Figure 3.27.

4. Finish the chamfer.

FIGURE 3.27 Applying a chamfer to the model edge

5. Rotate the part so you can see the bottom.

6. Select the sharp inside edge, and select the Chamfer tool.

7. Set the option to Two Distances (you may have to reselect the edge).

8. Set the first value to 2 and the second distance to 4.

9. Finish the chamfer. See Figure 3.28 for the finished adapter.

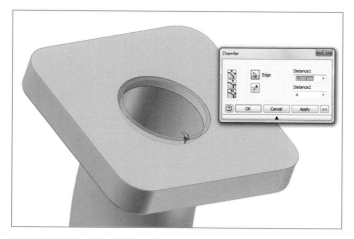

FIGURE 3.28 Adding the second chamfer with two different sizes

Now you can add the final touches to your part by placing holes.

Placing Concentric Holes

Hole features are critical geometry, and having the standards included in Inventor makes it easier to add them.

Certification
Objective

1. Make certain that the 2013 Essentials project file is active, and then open c03-15.ipt from the Parts\Chapter 3 folder.

2. Start the Hole tool from the marking menu or the Modify panel of the 3D Model tab.

3. In the Hole dialog box, set the placement to Concentric.

4. Select the top face of the bottom extrusion as the plane for the hole to begin.

5. Click the nearest 14 mm fillet face for the concentric reference.

6. Drag the hole's diameter by clicking and dragging the ring that appears on the face to see how this can be approximated.

 7. Set the hole type to Clearance for an ISO 24017 Hex Head Cap Screw with an M10 thread and Normal fit. See Figure 3.29.

8. Make sure the Termination setting is Through All.

9. Click OK to place the hole feature.

Using a Linear placement would use two edges to locate the hole but can also be used to "rough in" a hole by not setting a location.

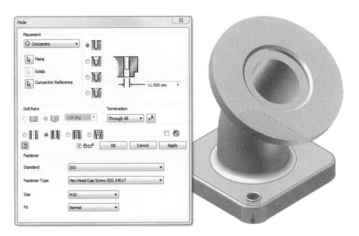

FIGURE 3.29 A clearance hole can size itself based on standards.

You can make the bolt pattern across the bottom plate in several ways. For instance, you can repeat the Hole command for each corner or do a rectangular pattern or since the first feature is centered on the z-axis of the part, you can use a circular pattern.

Creating a Circular Pattern

Certification Objective

Now you will make three more holes the easy way, and they will be associative to any change to the original hole:

1. Make certain that the 2013 Essentials project file is active, and then open the c03-16.ipt file.

 2. Select the Circular Pattern tool from the Pattern panel of the 3D Model tab.

3. In this case, you only need to select the Hole1 feature for the features selection.

4. In the dialog box, click the Rotation Axis button.

5. Expand the Origin folder in the browser if needed, and select Y Axis; you, can pick the lowest cylindrical face of the sweep feature.

6. Set the value in the Placement field to 4, and check the preview.

7. When the preview shows new holes added to the other corners, click OK to create them. Figure 3.30 shows the part at this point.

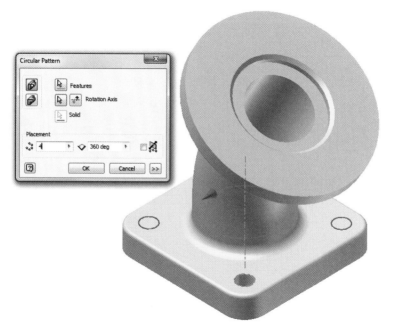

FIGURE 3.30 Using a pattern saves on repetition and helps prevent mistakes.

Now you can place the final holes on the part.

Placing Sketched Holes

To create complex patterns of holes, you can develop a sketch that includes a number of hole centers or naturally occurring points. By selecting them, you can place many holes in a single feature.

Certification Objective

1. Make certain that the 2013 Essentials project file is active, and then open c03-17.ipt from the Parts\Chapter 3 folder.

The dotted lines and circle that appear in the sketch have an override applied to them: Construction. This prevents them from being recognized as profiles by the 3D features but allows them to be used to construct other things. In this case, you'll use them for the location of the center points for holes. The corners of the square have an override applied as well. This will make the Hole tool find them automatically.

2. Start the Hole tool. As mentioned, four locations will be selected.

3. Set the type to Tapped Hole, the thread type to ISO Metric profile, and the size to 10. Select the Full Depth check box for the threads.

4. Change the termination to To, and click the lower side of the cylindrical face so the holes stop there. See Figure 3.31.

5. Select OK to complete the hole.

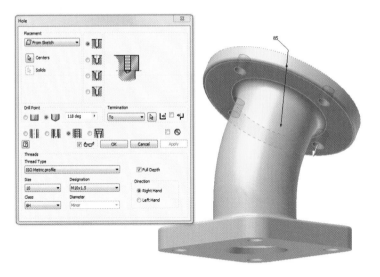

FIGURE 3.31 Hovering over an obscured object will eventually cause a drop-down to appear, allowing you to select it.

Now, your part is complete. You've used a lot of different features and a number of different techniques. You'll use even more features in Chapter 7 to create more complex geometry. The exact tools you will use depend on what it is you need to accomplish.

THE ESSENTIALS AND BEYOND

There are many more options in part modeling that will be covered in Chapter 7. In this chapter, you worked with the basic tools that will allow many people to do most of the things they need to do. You may also have noticed that the tools do a good job of prompting you for the things they need to help you.

ADDITIONAL EXERCISES

▶ Try modeling the base with the Polygon tool.

▶ Explore whether using an extruded circle is more appealing to you than using the Revolve tool.

▶ Try modifying the angular dimension on the sweep path sketch to see the effect.

▶ Experiment with different types of holes to see how they change with different fasteners.

CHAPTER 4

Putting Things in Place with Assemblies

Few people need to work on parts that exist on their own. Not only is working with assemblies necessary, but in many ways, Autodesk® Inventor® software works best in the assembly environment. To become really proficient with the Autodesk Inventor assembly environment, you must understand how to use assembly constraints and when to use which ones. Once you're familiar with the tools, it is best to think about how things fit together from a more elemental standpoint.

▶ **Creating an assembly**

▶ **Understanding grounded components**

▶ **Applying assembly constraints**

▶ **Working with the Content Center**

▶ **Using the Bolted Connection Generator**

▶ **Saving time with the Assemble tool**

Creating an Assembly

Certification Objective

To create an assembly, you follow the same basic process as beginning any other type of file in Autodesk Inventor. Starting a new assembly file changes the Ribbon to display tools that you need for the assembly process.

1. Open the New dialog box, navigate to the Metric template folder, and start a new assembly using the Standard (mm).iam template.

2. Select Place Component from the marking menu or the Component panel of the Assemble tab.

Place

3. Go to the Parts folder and then to Chapter 4; double-click the c04-01.ipt file and click OK.

   An instance of the component appears in the assembly immediately, and as you move your cursor, you will see that the part follows it, although it appears faded. This allows you to place multiple instances of the component in your assembly.

4. Press the Esc key to finish placing components.

5. Orbit your assembly so that you can see the Top, Front, and Right faces of the ViewCube.

6. Begin placing another component in the assembly, this time selecting the c04-02.ipt file from the same folder.

7. After placing one instance, end the command (see Figure 4.1 for reference).

8. Set this point of view to be your new Home view.

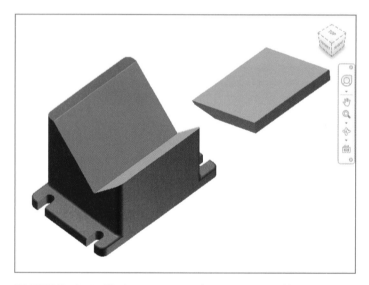

FIGURE 4.1 Placing components into a new assembly

If you know you need three instances of a component, you can also drag an instance from the Browser into the Design window to place another.

▶

Take a moment to look at your Browser. The components you placed appear just as if you'd placed views in a drawing or features in a part. You will see the name of the part displayed, followed by a colon and an additional number. This number denotes which instance of the component you are looking at.

Understanding Grounded Components

The first component placed in an assembly is referred to as the *base component* and, as in the real world, should be the component that the largest number of other components are attached to or rely upon for their foundation.

Certification
Objective

In the Browser, notice a thumbtack icon overlaying the cube icon that repre- sents the part. This thumbtack means the component is *grounded*, which means you don't need to constrain the base part to the coordinate system of the assem- bly file. Any or all components can be grounded, and the base component can be "ungrounded" if you want to be able to reorient it in the assembly.

Applying Assembly Constraints

The purpose of constraining components together in an assembly is to mimic the behavior of components in the real world. The constraints you apply do this by removing degrees of freedom from the components. Although it is not neces- sary to remove all the degrees of freedom, you will need to remove as many as needed to properly position the part. Inventor has relatively few constraint tools, but some of them can be used in many ways.

Certification
Objective

The Place Constraint dialog box has four tabs: Assembly, Motion, Transitional, and Constraint Set. Each accesses tools that offer different ways to solve your assembly. The selections area on each tab and with each type offers buttons with different colors that are reflected in the assembly as you select the entities. In this section, you'll use exercises to better understand the most commonly used constraint tools.

The Assembly tab (Figure 4.2) offers four types of assembly constraint: Mate, Insert, Angular, and Tangent. Each offers at least two solutions. In addition, the Transitional tab offers its own constraint. For the most part, you will see the options each constraint offers by looking at the icons for its solutions.

FIGURE 4.2 The Assembly tab of the Place Constraint dialog box

The Mate Constraint

This tool has two primary options: a Mate solution and a Flush solution. Each will reposition components based on the selected geometry with the ability to apply an Offset value in the dialog box that will create a gap or an interference between the component depending on whether the value is positive or negative and how the parts are aligned.

The Mate solution works on planes, axes, and points, and it behaves as though it puts these entities in opposition to one another. If you put a box on top of another box in the real world, you could say that the top of the lower box was pushing against the bottom of the upper one. In Inventor, this would be a Mate/Mate constraint.

First you will use the Mate solution in your assembly:

1. Make certain that the 2013 Essentials project file is active and then open c04-01.iam from the Assemblies\Chapter 4 folder.

2. Launch the Constrain tool from the Position panel or from the marking menu.

3. Set Constraint Type to Mate, and leave Solution at its default (Mate).

4. In the Selection group, the button for 1 will be depressed. Click the blue face on the red part for the first selection.

5. Once the first object is clicked, the second one will be enabled. Click the blue face on the purple part.

    The two parts automatically snap together (see Figure 4.3). You can turn this off by deselecting the Preview icon in the dialog box.

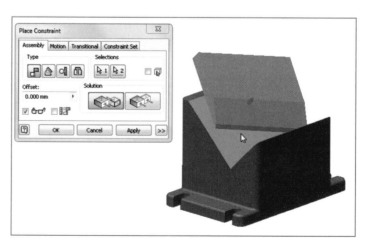

FIGURE 4.3 The components move into position as you apply constraints.

6. Click Apply in the dialog box or right-click and select Apply from the context menu to create the first constraint and prepare for another.

7. Orbit your assembly, and select the yellow face of the red part and the yellow face of the purple part.

8. Click OK to place the constraint and close the tool.

9. Drag the purple part in the assembly. You can see that it is still able to slide from side to side.

The Flush solution works only on planes and controls the alignment of faces to one another.

1. Verify that the 2013 Essentials project file is active and then open the c04-02.iam file from the Assemblies\Chapter 4 folder.

2. Launch the Constrain tool from the Position panel or the marking menu.

3. Set Constraint Type to Mate, and set Solution to Flush.

4. Select the sides of the two components, as shown in Figure 4.4.

5. With Preview on, the components will align. Click OK to create the constraint.

FIGURE 4.4 The Flush constraint aligns faces.

The simple Mate constraint and its solutions cover the needs of many Inventor users. To better understand which constraints are missing in an assembly, you can use the Degrees Of Freedom (DOF) tool.

Certification
Objective

DEGREES OF FREEDOM

As mentioned earlier in this chapter, Autodesk Inventor uses the concept of degrees of freedom to describe what movement can be realized by a component. Unless a component is grounded or constrained, it has six degrees of freedom—three rotational and three translational around and along the x-, y-, and z-axes. As you apply constraints, the ability to move in particular directions is removed.

The Insert Constraint

The Insert constraint is a hybrid and combines a Mate constraint on the axis and between the adjacent faces of curved edges. It has many uses, but it is most commonly used to put round parts in round holes.

1. Verify that the 2013 Essentials project file is active and then open the c04-03.iam file from the Assemblies\Chapter 4 folder.

2. Switch the Ribbon to the View tab.

3. Click the Degrees Of Freedom tool in the Visibility panel.
   An icon appears on the purple part, showing that it has all six degrees of freedom.

4. Launch the Constrain tool from the Assemble tab in the Position panel or the marking menu.

5. Set Constraint Type to Insert, and leave Solution set to Opposed.

6. Select the circular edge of the hole on the purple part's flat face, and then select the circular edge of the hole on the red part, as shown in Figure 4.5.

7. When the parts are aligned, click OK.
   The DOF icon changes to show one remaining rotation of freedom.

8. Drag the purple part to see how it moves.

Being able to exercise the remaining degrees of freedom is not limited to a single component. You can use this to show how a mechanism works.

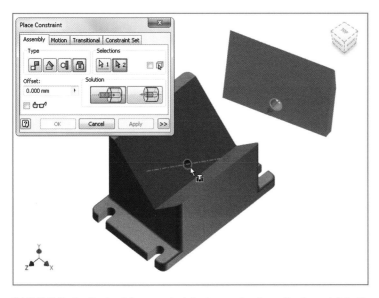

FIGURE 4.5 Applying constraints does not automatically restrict all movement.

The Angular Constraint

There are three solutions for the Angular constraint, allowing the flexibility to, for example, set the axis of rotation apart from the components, along with other options that give less control. For this exercise, you will use the most precise solution, the Explicit Reference Vector tool.

1. Verify that the 2013 Essentials project file is active, and then open the c04-04.iam file from the Assemblies\Chapter 4 folder.

2. Launch the Constrain tool from the Assemble tab in the Position panel or the marking menu.

3. Set Constraint Type to Angle, and make sure the solution type is set to Explicit Reference Vector.

4. For the first selection, click the narrow face on the side of the purple part.

5. For the second, click the face on the red part, as shown in Figure 4.6.

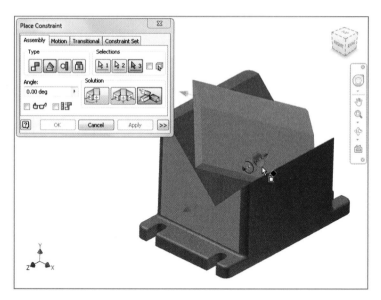

FIGURE 4.6 Complexity in a constraint also adds reliability and stability.

6. For the third, click the cylindrical face of the hole in the purple part.

7. Click OK to complete constraining of the purple part. The purple part can no longer rotate freely around the axis of the hole. This constraint is a great option for aligning faces that aren't flush on all sides.

A good use for an Inside solution could be a pin following a curved face. An Outside could be rollers in contact with each other.

The Tangent Constraint

When rounded faces need to stay in contact with other rounded faces, the Tangent constraint is often the only tool you will need. The constraint can have Inside or Outside solutions.

1. Verify that the 2013 Essentials project file is active and then open the c04-05.iam file from the Assemblies\Chapter 4 folder.

2. Launch the Constrain tool from the Assemble tab in the Position panel or the marking menu.

3. Set Constraint Type to Tangent, and make sure the solution type is set to Outside.

4. Make your first selection, the blue, curved face on the orange part.

5. Make the second selection, the blue face on the gray part.

6. Click OK to bring the parts together.

7. Drag the orange part to see that the selected faces stay in contact.

The tangency added follows the entire cylindrical face. If the face isn't a full cylinder, components constrained to it will follow the curvature of the face even if it moves off the end of the face.

The Transitional Constraint

A fifth type of constraint is available on its own tab. Where the Assembly tab's Tangent constraint works well for simple curved faces, the Transitional constraint works best with changing faces such as cams.

1. Verify that the 2013 Essentials project file is active, and then open the c04-06.iam file from the Assemblies\Chapter 4 folder.

2. Launch the Constrain tool from the Assemble tab in the Position panel or the marking menu.

3. In the Place Constraint dialog box, switch the tab to Transitional.

4. Click the yellow face on the orange part for the first selection.

5. Click the yellow face on the gray part for the second.

6. Click OK to accept the constraint. See Figure 4.7.

7. Drag the orange part to see how the Transitional and Tangent constraints work together to position the two components.

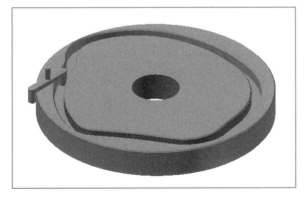

FIGURE 4.7 Different types of constraints can be combined to create useful models.

Adding a Transitional constraint offers more flexibility on the type of faces to which you can constrain than the Tangent constraint. Now, you will take a look at working with other data sources in the assembly.

Working with the Content Center

During installation, you have the option of installing various types of standard components. Depending on the options you choose, you will have tens or hundreds of thousands of standard components available to you. Rather than creating part files for bolts, nuts, and so on, you can select them from a library.

1. Verify that the 2013 Essentials project file is active, and then open the c04-07.iam file from the Assemblies\Chapter 4 folder.

2. Locate the Place From Content Center tool under the Place tool in the Component panel of the Assemble tab.

   This will launch the Place From Content Center (Figure 4.8) dialog box, where you can locate many kinds of standard components.

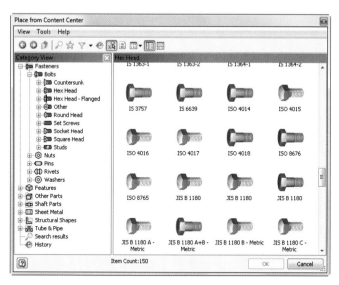

F I G U R E 4 . 8 Hundreds of thousands of standard components are built into Inventor.

3. Expand the categories by selecting Fasteners ➤ Bolts ➤ Hex Head.

4. Double-click ISO 4017.

When the dialog box disappears, a small preview of the fastener will appear. This preview is one of the available sizes for the bolt.

5. Hover over the edge of the hole in the purple part until the fastener resizes based on the hole's size.

6. After the bolt resizes, click when the edge of the hole highlights in red (Figure 4.9).

FIGURE 4.9 The fastener will resize based on an existing hole.

This temporarily places a detailed preview the bolt in the hole.

7. Without ending the Constraint tool, change the Ribbon to the View tab, and change Visual Style to Shaded With Hidden Edges.

8. Click the Front face of the ViewCube to change the view.

9. Click and drag the red arrow at the tip of the bolt until it shows that the fastener size is at M10 35, and then release to set the length (Figure 4.10).

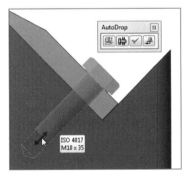

FIGURE 4.10 Fastener lengths can be changed in standard sizes.

10. Click the check mark in the AutoDrop mini-dialog box to place the fastener in the assembly.

11. The tool will place the fastener and then offer to place another bolt. Press Esc to finish the command.

12. Switch to the Home view, and move the fastener to see that it is constrained to the hole.

The remaining degree of freedom on the fastener does not need to be removed. As long as the fastener is seated and aligned with the hole, it is effective in the assembly.

Using the Bolted Connection Generator

Certification
Objective

The process of applying holes to components and then positioning fasteners is not difficult, but it does put the focus on the features of the software rather than the needs of the design.

The Bolted Connection Generator tool will allow you to put standard components into an assembly and use engineering calculations to be sure you are using the correct component.

1. Verify that the 2013 Essentials project file is active, and then open the c04-08.iam file from the Assemblies\Chapter 4 folder.

2. Activate the Design tab on the Ribbon.

3. Start the Bolted Connection tool.
   The Bolted Connection Component Generator dialog box (Figure 4.11) has a number of options, but it works very much like the Hole feature in a part model.

4. In the Type group, set the hole type to Blind Connection Type.

5. For the start plane, use the yellow face of the purple part.

6. For Linear Edge 1, click the left edge of the yellow face. If you are prompted for a value for this dimension, use 50 mm, as shown in Figure 4.12.

7. Click the bottom edge for Linear Edge 2. If you are prompted for a value for this dimension, use 15 mm for the value.

8. Click the face where the two parts are in contact to select the Blind Start Plane. You will be able to pick through the purple part.

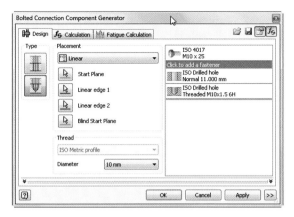

FIGURE 4.11 The Bolted Connection Component Generator dialog box

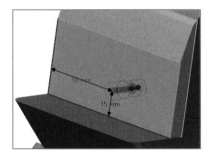

FIGURE 4.12 The dimensions locate the placement of the bolted connection.

9. In the Thread group, set the Thread class to ISO Metric Profile.

10. Set the diameter to 10 mm.

11. Click where it says *Click to add a fastener* in the column on the right. After a short time, a list of bolts that will fit the hole as currently defined appears. You can filter the list to see only those bolts of a single standard.

12. In the component selection pop-up, change Standard to ISO to filter the options. See Figure 4.13.

13. Click ISO 4017.

14. Change the view in the Design window to the Front view of the assembly.

15. Drag the red arrow at the end of the bolt until it shows you are placing an M10 35 bolt (Figure 4.14).

16. Click OK to finish the tool and OK again to approve the creation of a file.

17. Double-click the red part to see that a tapped hole has been added to it.

18. Click the Return tool at the end of the Ribbon to return to the assembly.

FIGURE 4.13 Filtering fasteners based on standards makes selection easier.

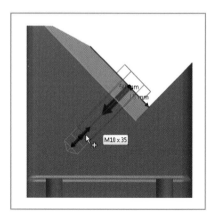

FIGURE 4.14 In a bolted connection, you can drag the hole depth and bolt length.

Placing fasteners using this tool saves time over placing holes in advance and using the Content Center. If the assembly changes, you might have to edit the bolted connection and click OK to update it. I will review Design Accelerator tools in Chapter 8.

Saving Time with the Assemble Tool

The Assemble tool works as a shell for many of the constraint tools. Based on the geometry you select, it will preview a constraint and allow you to select a different option or enter an offset or angle value using a mini-toolbar near the cursor.

1. Verify that the 2013 Essentials project file is active, and then open c04-09.iam from the Assemblies\Chapter 4 folder.

2. Start the Assemble tool from the Position panel of the Assemble tab.

3. Click the blue face of the purple part and then the blue face of the red part.

4. Click the Apply icon in the mini-toolbar to add this Mate/Mate constraint, and move to the next constraint.

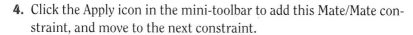

5. Now select the yellow face on the purple part and the yellow face on the red part. Apply the constraint.

6. Now click the faces on the left of each part, and notice that the mini-toolbar shows that it will add a Mate/Flush constraint.

7. Click the OK button on the mini-toolbar to finish using the Assemble tool on the purple part.

8. Press the Enter key or the spacebar to start the Assemble tool again.

9. Click the underside of the bolt where the shaft meets the head.

10. Click the hole in the purple part, and click OK to place an Insert constraint between the bolt and the hole feature (Figure 4.15).

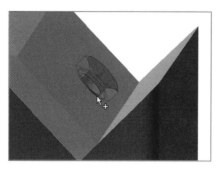

FIGURE 4.15 Clicking the second edge previews how the component will be repositioned.

The Assemble tool works well, and some experienced users are beginning to prefer it to using the traditional Place Constraint dialog box. A couple of things to keep in mind are that you should not select a grounded component and you must restart the Assemble tool when you want to apply constraints to different parts.

THE ESSENTIALS AND BEYOND

This chapter was designed to cover the basics of the assembly process of Inventor. Because of the simplicity of assemblies in Inventor, I covered the most commonly used tools and even some more advanced options.

ADDITIONAL EXERCISES

▶ Review the components in the Content Center to see the extent of the options.

▶ Experiment with the Assemble tool for doing all of the exercises.

▶ Work with different fasteners and standards in the Bolted Connection Component Generator dialog box.

CHAPTER 5

Customizing Styles and Templates

Consistent standards improve communication and quality. Autodesk® Inventor® software comes with several standards from around the world that can be used as a foundation for constructing a company-specific standard. Inventor standards are based on a collection of styles. In this chapter, you will learn to modify styles and come to understand the process.

Another easy way to create consistency among users is to use templates with preconfigured styles. These templates can even be configured to short-cut the process of defining a part, assembly, or drawing.

▶ **Working with styles**

▶ **Defining a new material**

▶ **Defining a title block**

▶ **Saving a new template**

▶ **Creating a quick-start template**

Working with Styles

A good strategy for making changes to styles is to start small and work your way up. Some style elements are shared as a resource for others. By establishing the properties of these resources, you can simplify configuring others above them.

Creating a Standard

You'll begin the process by creating a new standard, the specific settings of which will be defined in following exercises:

1. Verify that the 2013 Essentials project file is active, and then open the file c05-01.idw from the Drawings\Chapter 5 folder.

2. Switch the Ribbon to the Manage tab, and from the Styles And Standards panel, launch the Style and Standard Editor dialog box (Figure 5.1).

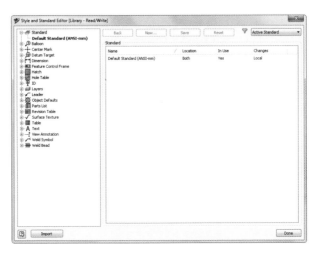

FIGURE 5.1 The Style and Standard Editor dialog box controls all aspects of element display in a drawing.

3. In the top right of the dialog box, set the drop-down to Local Styles. The column on the left of the dialog box has the various categories of styles.

4. At the top of that column, right-click the Default Standard (ANSI-mm) standard, and select New Style from the context menu.

5. In the dialog box, enter **Essentials (in)** as the name, and leave ANSI as the standard it is based on.

6. Click OK to create the standard.

7. In the dialog box, double-click the new standard to make it active.

It is good practice to create your own style as you've begun doing, rather than just modifying an existing standard or style. That way, you can maintain the original, and it will allow more flexibility for developing templates without generating error messages.

Creating a New Object Default Set

Instead of lines, arcs, and circles, Inventor drawings consist of objects. A visible line is a different type of object than a hidden line, and therefore, it can be controlled independently. A standard refers to the Object Default Set to know how to display the elements of the drawing.

1. Continue using the open file, or verify that the 2013 Essentials project file is active and then open c05-02.idw from the Drawings\Chapter 5 folder.

2. Start the Style and Standard Editor from the Styles And Standards panel of the Manage tab.

3. Expand the Object Defaults category in the column on the left.

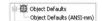

4. Right-click the Object Defaults (ANSI-mm) style, and create a new style from the context menu.

5. Enter **Essentials (in)** for the name of the new style, and click OK to create it.

Creating a new Object Default Set is optional when creating a new standard but makes it easier to keep standards unique. If you share an Object Default Set between standards, a change to it is reflected in all of them. The same will be true for many of the modifications made in this chapter, which are also essential stages in building a standard.

Defining a Text Style for Dimensions and Notes

The text style is a perfect example of a style you create once that is used by more than one object type.

1. Continue using the open file, or verify that the 2013 Essentials project file is active and then open c05-03.idw from the Drawings\Chapter 5 folder.

2. Start the Styles and Standards Editor from the Styles And Standards panel of the Manage tab.

3. Expand the text styles, right-click the Note Text (ANSI) style, and select New Style from the context menu.

4. Enter **Essentials Text** for the new text style name, and click OK to add it to the standard.

5. On the right side of the dialog box, select Arial from the Font drop-down menu.

6. In the Paragraph Settings area, click the Color (By Layer) icon to open the Color dialog box.

   Black is already selected in the dialog box, but in the upper left of the dialog box, the By Layer override is selected. This will change the color of the text to match whatever layer it is set to.

7. In the Color dialog box, deselect the By Layer override, and click OK to set the color of the text to black. See Figure 5.2.

8. Click the Save icon at the top of the dialog box to save the new standard in the drawing.

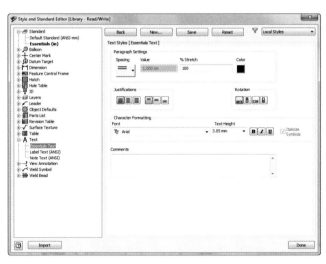

FIGURE 5.2 Changing the properties of the text used in dimensions and notes

Now that you have a new text style, let's put it to work.

Defining a New Dimension Style

There are many opinions on the proper way to display dimensions even in relatively well-defined standards like ISO. Now that you have your text style, you can practice creating a simple change to define a new dimension style.

1. Continue using the open file, or verify that the 2013 Essentials project file is active and then open c05-04.idw from the Drawings\Chapter 5 folder.

2. Start the Style and Standards Editor from the Styles And Standards panel of the Manage tab.

3. Expand the dimension styles, right-click the Default (ANSI) style, and select New Style from the context menu.

4. Enter **Essentials (in)** for the new dimension style name, and click OK to add it to the standard.

5. With the Essentials (in) standard highlighted, click the Text tab on the right side of the dialog box.

6. Change Primary Text Style to the Essentials Text style using the drop-down menu.

7. In the Orientation group of the Text tab, change the first Linear option to display the text over the dimension line.

8. Change the Diameter and Radius options to display the dimension value in line with the leaders, as shown in Figure 5.3.

9. Click the Save icon at the top of the dialog box to save the new standard in the drawing.

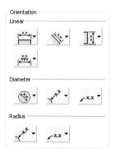

FIGURE 5.3 Changes to style options are represented in the icons.

Now, you can change some of the appearance settings that will have an effect on the other changes you've made.

Setting Layer Properties

Layer properties can be set from the Annotation tab in the drawing, but establishing them in the Style and Standards dialog box is a good place, too.

1. Continue using the open file, or verify that the 2013 Essentials project file is active and then open the file c05-05.idw from the Drawings\ Chapter 5 folder.

2. Start the Style and Standards Editor from the Styles And Standards panel of the Manage tab.

3. Expand the Layers styles, and click the Dimension (ANSI) layer to see its properties and the properties of the other layers in the dialog box window on the right.

   You can change the layer properties displayed by clicking their current value and using a drop-down or a dialog box to change them. You can also click the lightbulb icon to change the visibility of the layer.

4. Click the black rectangle in the Dimension (ANSI) row to open the Color dialog box.

5. Click the dark blue swatch in the top row of the Color dialog box, and click OK to set the color of the Dimension layer.

6. Change the color of the Hatch (ANSI) layer to the lightest gray on the bottom row.

7. Change the color of the Hidden (ANSI) layer to the light red on the top row.

You can also define custom colors in the dialog box.

8. Click the Save icon at the top of the dialog box to save the changes to the layers in the drawing.

Note that the changes to the layer styles have updated the drawing appearance.

Setting the Object Defaults

As mentioned earlier in the chapter, the Object Default Set controls the elements that determine the appearance of the elements in the drawing. The preceding exercise changed the colors of the layers that objects were on, but the Object Default Set determines the default layer and other styles that the objects use in the first place.

1. Continue using the open file, or verify that the 2013 Essentials project file is active and then open c05-06.idw from the Drawings\Chapter 5 folder.

2. Start the Style and Standards Editor from the Styles And Standards panel of the Manage tab.

3. Return to the Object Defaults style, and select the Essentials (in) style to display its setting on the right.

4. Click the Object Style column header twice to sort the styles alphabetically.

5. Scroll down using your wheel button or the slider bar on the right until you see the various dimension types that use the Default—mm (ANSI) dimension style.
   Clicking an object style or layer on the right will display a drop-down list of options.

6. Change all of the objects using the Default—mm (ANSI) style to the Essentials (in) style using the drop-down menu.

7. Save the changes by clicking the Save icon at the top of the dialog box.

8. Change the Filter drop-down above the Object Type column to Model/View Objects.

9. Locate the objects using the Hidden Narrow (ANSI) layer, and change them to the Hidden (ANSI) layer.

10. Click the Save icon at the top of the dialog box to save the changes to the Object Default Set.

11. Click Done to close the dialog box.

> ◀
>
> To speed things up, you can click a drop-down and press the key of the first letter to jump to that value. In this case, you can press E.

Now, you have control over the object that will be added to your drawing going forward. You can update the existing objects or even publish these changes to be used in other drawings or templates in the future.

Saving the Standard

The standard as you've defined it so far is ready to use as a starting point for further style definitions. Later, there may be elements that you change in a drawing that you may not want to reuse in others, or you may want to change certain properties in individual templates rather than globally changing the standard. So, this is a good time to save the standard in its current form.

1. Continue using the open file, or verify that the 2013 Essentials project file is active and then open c05-07.idw from the Drawings\Chapter 5 folder.

2. Click the Save tool in the Styles And Standards panel of the Manage tab.

3. In the Save Styles to Style Library dialog box, set the value in the Save To Library? column to Yes only for the items you created. See Figure 5.4.

4. Click the Cancel button in the dialog box.

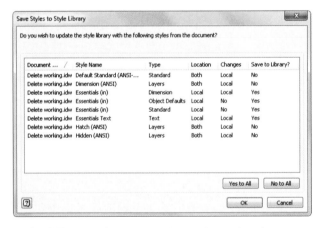

FIGURE 5.4 Saving the unique styles so that they can be reused

The previous exercise shows you that you can still choose what changes get saved to the standard. Some of the changes will affect objects that are used in standard templates, and saving the changes made to the layers at this point may cause conflicts with the default templates. It is best for you to make these changes to your templates with your own preferences when the time comes.

Changing a Color Style

Styles exist in other file types as well. In part and assembly files, there are styles for the colors and materials and even lighting. Making changes to these styles is easy and a little more direct than changing the styles in a drawing.

1. Verify that the 2013 Essentials project file is active, and then open c05-02.ipt from the Parts\Chapter 5 folder.

2. Start the Style and Standards Editor from the Styles And Standards panel of the Manage tab.

3. Expand the Color category in the column on the left.
   Looking at the list, you will see that the active color is in bold. This color may be displayed in the Quick Access toolbar if it is set to something other than As Material.

4. Right-click the current color, Blue Pastel, and select New Style from the context menu.

5. Set the new style name to **Essentials Blue**, and click OK to save it in the current file.

6. Double-click the new style to make it the active style for the part.
   On the Basic tab of the dialog box, you can change the color properties to affect the appearance of the part. The Texture tab allows you to select an image file that will add a visual texture to the part. The Bump Map tab will allow you to apply a texture that will appear to alter the surface of the part.

7. In the Colors group on the Basic tab, click the Ambient icon to open the Color dialog box.

8. Near the bottom of the Color dialog box, click the Define Custom Colors icon to expand the color selection options.

9. Below the words *Custom colors* is a series of empty color fields. Click one to begin editing it.

10. In the lower right of the dialog box, enter the following values for your new color:

   > Red: 60
   >
   > Green: 80
   >
   > Blue: 140

11. Click the Add To Custom Colors button to make the changes to the selected color.

12. Click OK to change the ambient color of the part. See Figure 5.5.

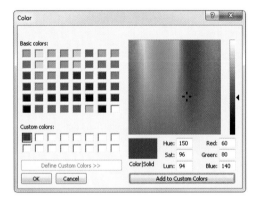

FIGURE 5.5 Defining a custom color for parts

13. Back on the Basic tab, set the Shininess level to 60%.

14. Click Save at the top of the dialog box and then Done at the bottom to close it.

The results may not be earth-shattering, but the ability to customize the colors can really add a nice touch for visualizing your designs.

Defining a New Material

In the iProperties section of the Physical tab, when you change the material of a part, it may change color as well, simply because a color is associated with a particular material. The two are, in fact, completely separate ideas.

The Material value is what the mass and inertial properties of the part are based upon, and although Inventor has many materials built in, it is common for users to want to add a specific one for their needs.

1. Continue using the open file, or verify that the 2013 Essentials project file is active and then open c05-03.ipt from the Parts\Chapter 5 folder.

2. Start the Style and Standards Editor from the Styles And Standards panel of the Manage tab.

3. Expand the Material category in the column on the left.

4. Double-click Aluminum-6061.

5. Review the Material properties on the right, changing the Default Units drop-down to see the updated values.

6. Close the dialog box by clicking the Done button at the bottom.

You can right-click a material similar to one that you want to add to the library and select New Style to get a head start. After you've adjusted the appropriate material properties, you can save the material in the part, save it into the standard, or export it to be used by others.

Saving changes made to the choice of materials to your standard is a good way to ensure accuracy and quality and is highly recommended.

In addition to existing corporate and vendor material specifications, there are many online resources for getting information about materials.

Defining a Title Block

A common need for new users of Inventor is to modify the title blocks of their drawing templates to match their existing drawings. The process is relatively simple but not obvious. You will make a simple change, but the workflow is the same regardless of complexity.

Certification
Objective

A company logo or any image in .bmp, .gif, .jpg, .png, or .tif format can be added to a title block. The image can be embedded or linked.

1. Verify that the 2013 Essentials project file is active, and then open c05-08.idw from the Drawings\Chapter 5 folder.

2. In the Browser, locate and expand the Drawing Resources folder, and then expand the Title Blocks folder.

3. Right-click the ANSI-Large title block, and select Edit from the context menu.

Opening a title block for editing shows you something that looks like a disaster (see Figure 5.6). As you look closer, you will see that all the dimensions are there to control the location of lines, text, and properties that automatically populate as you begin to place views of components in the drawing.

4. Zoom in on the lower-left corner of the title block. Make sure you can clearly see the space with the word *Mass* in the upper-left corner.

5. Select the property text *<ENG APPROVED BY>* above the Mass cell.

6. Press Ctrl+C to copy the property to the Windows Clipboard.

7. Press Ctrl+V to paste a copy back into the title block.
   The copied property will usually be duplicated just above the original text.

8. Click and drag the new text down to the open cell below the word *Mass*.

9. Use a Coincident constraint to connect the origin point of the text to the point at the intersection of the title block, as shown in Figure 5.7.

10. Once the text moves, use the Esc key to stop the constraint.

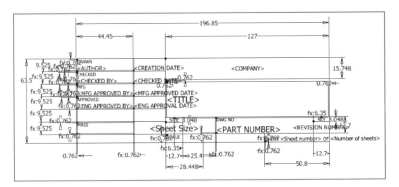

FIGURE 5.6 The line work, text, and properties behind a title block

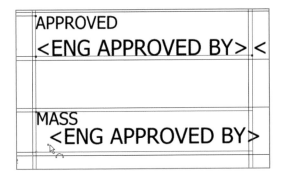

FIGURE 5.7 Sketch constraints work to control the location of text in the title block.

11. Right-click the property and select Edit Text from the context menu.

12. Remove the existing text from the bottom window of the dialog box.

13. In the Format Text dialog box, change the Type drop-down to Physical Properties—Model and the Property drop-down to Mass.

14. Click the Add Text Parameter button to update the property below.

15. The new property will appear in the Text window. See Figure 5.8.

FIGURE 5.8 Text can be created automatically based on model properties.

16. Click OK to update the title block.

17. Select Save Title Block from the marking menu.

18. Click Save As from the dialog box that appears to create a new title block using your changes.
    Another dialog box opens and prompts you for the name of the new title block.

19. Enter **Essentials** for the new title block, and click Save.

20. Under Drawing Resources ➢ Title Blocks in the Browser, double-click the Essentials title block to place it in the drawing.

You've now added a new resource to the drawing that can be shared with others or used to define a new personalized template for future drawings.

Saving a New Template

Now, you can create a template based on the existing drawing. It is really a very easy thing to do.

1. Verify that the 2013 Essentials project file is active, and then open c05-09.idw from the Drawings\Chapter 5 folder.

2. Under the Application menu, expand the Save As options, and select Save Copy As Template.

3. In the Save Copy As Template dialog box, click the Create New Folder icon.

4. Name the new folder Essentials, and double-click to open it.

5. Give your new template the filename of Essentials 1, and click the Save button.

6. Start the New File tool from the Quick Access toolbar, and note that there is now a folder in the dialog box named Essentials.

7. Select the folder to see your new template ready to be used and is categorized based on the file type. See Figure 5.9.

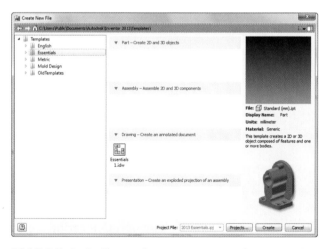

F I G U R E   5 . 9 If users share a common template source, it is easy to rebuild existing templates.

Now, you've created a new drawing template, and you can create as many different templates as you like. You can even open existing AutoCAD templates and make them Inventor templates.

Creating a Quick-Start Template

If you create components of similar sizes on a regular basis and have a consistent format for your drawings, you can configure a template to shortcut the drawing view placement process.

You can also create part or assembly templates with different default colors or even some features or components already placed.

1. Verify that the 2013 Essentials project file is active, and then open c05-10.idw from the Drawings\Chapter 5 folder.

2. Under the Application menu, expand the Save As options, and select Save Copy As Template.

3. Navigate to the Essentials folder if you did the previous exercise. If you did not, please review steps 3 through 5 of that exercise.

4. Save the new drawing template as Essentials Six View.

5. When you have returned to the Autodesk Inventor Design window, start the New File tool.

6. Double-click the Essentials Six View template to create a new drawing. This will open the Select Component dialog box, where you will select the file from which you want to create drawings.

7. Click c05-04.ipt from the Parts\Chapter 5 folder, and then click OK to close the dialog box.

This will generate a new drawing with the same scale, spacing, and positions as the drawing from which you built the template. Comparing the new drawing to the c05-10.idw file will allow you to see the difference more clearly.

THE ESSENTIALS AND BEYOND

Changes that you make to the standards and their various styles are limited only by you or your company's existing guidelines. Careful consideration of strategies for templates can save many hours if done correctly.

ADDITIONAL EXERCISES

▶ Open an existing DWG-based template only, and set up the Object Defaults Set to use its layering and other features.

▶ Try taking a part and creating a template from it that you could use as a "starter part."

▶ Seek out a specialized material that you use and add it to the Style library.

▶ See what other types of views will work in a preconfigured, multiple-view template.

CHAPTER 6

Creating Advanced Drawings and Detailing

In Chapter 2, "Creating 2D Drawings from 3D Data," you learned how to create the basic drawing views and what some might consider advanced drawing views of a part file. Placing views of assemblies is not any different from an individual part file. In this chapter, I will expand on those lessons, and you will begin to work with very advanced drawing creation and editing tools. You will also explore additional dimensioning and annotation tools.

▶ **Creating advanced drawing views**

▶ **Using advanced drawing annotation tools**

Creating Advanced Drawing Views

The part modeling process normally combines the uses of sketched and placed features. In Chapter 2, you created a section view that was based on a sketch. Projected views are placed by projecting from a parent view. Advanced drawing views often use sketches for their definition and can even use more than one sketch. Even a projected view utilizes advanced technologies to help make decisions on your behalf.

Projecting Views from a Section View

Certification Objective

You can place as many base views in a drawing as you want, but if you already have a section view with the proper orientation, you can use it as a parent view.

1. Verify that the 2013 Essentials project file is active, and then open the c06-01.idw file from the Drawings\Chapter 6 folder.

2. Right-click the section view, select Projected View from the marking menu, and drag the new view above and to the right of it.

An isometric view of the geometry that is visible in the section view is created.

Projected

3. Click the page to place the view; then right-click and select Create from the context menu to generate the view, as shown in Figure 6.1.

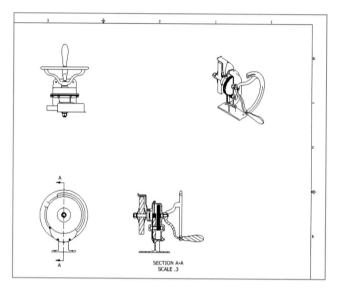

FIGURE 6.1 Placing an isometric view of a section view

4. Right-click the section view again, and select the Projected View tool from the marking menu.

5. Place the view to the right of the section view, and create the new view. Figure 6.2 shows the result.

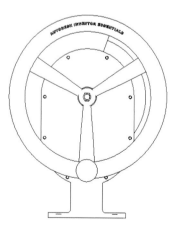

FIGURE 6.2 A complete view can be projected from a section view.

The new view is not a partial view; it is a full view. Because the new view is an orthographic projection, all of the geometry is included.

Creating a Sketch in the Drawing View

There are many uses for creating sketches in a drawing view. In this case, you need to create geometry to modify the drawing view:

1. Verify that the 2013 Essentials project file is active, and then open c06-02.idw from the Drawings\Chapter 6 folder.

2. When you hover over the bottom-left view, the boundary will highlight. When the view highlights, click the view, and then select Create Sketch from the Sketch panel on the Place Views tab.

3. When the Sketch tab appears, draw a spline in the view similar to the one in Figure 6.3.

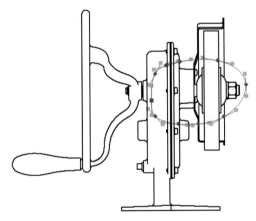

FIGURE 6.3 Define a sketch in the drawing view to make changes to the view.

4. After the loop of the spline is closed, right-click and select Finish Sketch on the marking menu to finish the sketch.

   It is important to verify that the sketch is associated with the drawing view.

5. To verify association, drag the drawing view, and make sure the sketch moves with it.

If the sketch is not attached to the drawing view, delete the sketch you made, and go through the steps again to make sure that the view is highlighted when you start the Create Sketch tool.

The Break Out View

The sketch you created will be used as a boundary to remove part of the components of the assembly so you can see the interior.

Break Out

1. Verify that the 2013 Essentials project file is active, and then open c06-03.idw from the Drawings\Chapter 6 folder.

2. Click the Break Out tool in the Modify panel of the Place Views tab.

3. Select the left view, and when the Break Out dialog box opens, you will notice it has already selected the spline since it was the only closed profile.

   You will then need to select the depth to which you cut through the parts.

▶

There are several options for selecting what components will be cut using the Break Out tool.

4. Hover over a curved edge. When the green dot appears, click it, as shown in Figure 6.4.

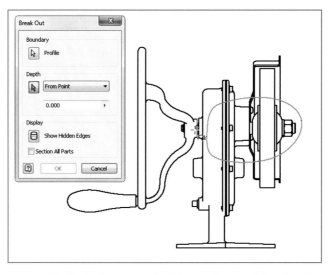

F I G U R E   6 . 4 After you select the profile, you determine the depth of the cut.

5. Click OK to generate the view.

This generates and cuts any and all parts down to the selected point, but the traditional line view drawing still makes it difficult to distinguish the components.

Changing Part Drawing Behavior

In this exercise, you will modify how the Autodesk® Inventor® software processes generating the drawing view of certain parts in the assembly:

1. Verify that the 2013 Essentials project file is active, and then open c06-04.idw from the Drawings\Chapter 6 folder.

2. Double-click the break out view to edit it.

3. Deselect the Style From Base check box.

4. Select the Shaded icon, and click OK to update the view.

In Figure 6.5, you can see there are a couple of errors. The first is that the screw and washer appear to be floating in the middle of the view. The second is a gear that is sectioned because of the way it was created. The third item flagged is not a mistake, but I want to show the seal on the shaft sectioned.

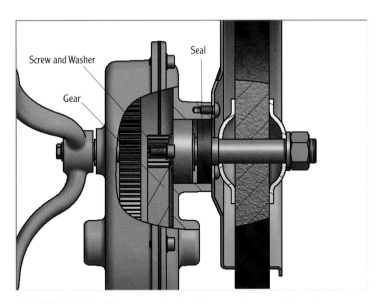

FIGURE 6.5 You can control whether parts are sectioned or even visible in a drawing view.

When you move your cursor over the view, all the edges highlight as you pass over them. By default, Inventor uses an Edge priority for selecting geometry in a drawing. You want Inventor to highlight the parts as you pass over them instead. This will make selecting them much easier.

5. Find the Selection drop-down on the Quick Access toolbar, and change the current selection priority to Part Priority by selecting it from the list.

6. Hold the Ctrl key to select the screw and the washer. Right-click the edge of one of the parts, and then select Section Participation from the context menu.

7. When the menu expands, select Section.

   The view regenerates with the screw and washer affected by the break out section, and therefore, they disappear.

8. Select the gear, and choose None from the Section Participation options.

9. Select the seal, and then choose Section.

This will again update the view to show the seal sectioned around the shaft. These changes will each update the view to appear like Figure 6.6.

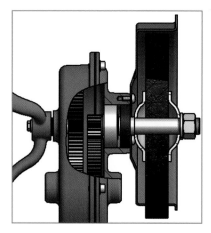

FIGURE 6.6 The updated view with the updated part visibility options

Part Visibility

If the screw and washer in the previous exercise weren't in a section or break out view, it would be possible to just remove the visibility. To do this, you must break the association between the drawing view and the design view of the assembly using the view edit dialog box. This has been done for you in the file referenced in the next exercise.

1. Verify that the 2013 Essentials project file is active, and then open c06-05.idw from the Drawings\Chapter 6 folder.

2. From the Quick Access toolbar, set the Selection filter to Part Priority.

3. Hover over the shroud. When it highlights, right-click and deselect Visibility.

The drawing view regenerates the previously hidden edges that are now visible (Figure 6.7).

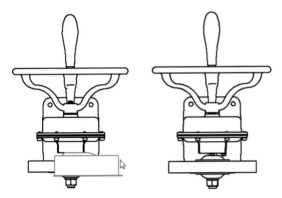

FIGURE 6.7 Selecting a part and turning it off in the view

View Suppression

A drawing view can be a stepping-stone to generate another, more important view. Being able to remove the clutter of the parent view can be useful.

1. Verify that the 2013 Essentials project file is active, and then open c06-06.idw from the Drawings\Chapter 6 folder.
   View1 is the base view in the drawing. It is located in the upper-left corner.

2. Find View1:c06-09.ipt in the browser, hover over it, and when the view highlights, right-click.

3. Select Suppress from the context menu.

Even though View1 is the base view for all the other views, you removed it without disruption to the rest of the drawing.

Drawing Element Suppression

You can suppress individual lines and other elements of the drawing view:

1. Verify that the 2013 Essentials project file is active, and then open c06-07.idw from the Drawings\Chapter 6 folder.

2. Zoom in on the isometric view in the upper-right corner.

3. Click and drag a window from left to right around the entire view.

4. When the lines in the drawing highlight, right-click and deselect Visibility from the context menu. Figure 6.8 shows the result.

FIGURE 6.8 A view can have visible edges removed for a more pictorial look.

Once an edge has been suppressed, it will remain that way even if the geometry changes size. New geometry added to the components will appear in the drawing view, so it is good practice to monitor suppressed edges.

 TIP   You can restore edges that you've made invisible by right-clicking the view and selecting Show Hidden edges from the context menu. The hidden edges will highlight, and you can pick the ones you want visible again. When you're finished, select Done from the context menu or even Show All.

Break View

When a part is long or has a large portion that consists of the same geometry, a *break* view can remove a portion of the drawing to make it easier to detail the dissimilar portions.

1. Verify that the 2013 Essentials project file is active, and then open c06-08.idw from the Drawings\Chapter 6 folder.

2. Zoom out until you can see the entire drawing view as it extends beyond the border of the drawing.

3. On the Place Views tab, locate and start the Break tool in the Modify panel.

Break

4. Click the Drawing view to open the Break dialog box.

5. Set the Gap value to .5.

6. Click the first point of the break just to the right of the red part and the second just to the left of the threaded end, as shown in Figure 6.9.

7. Add a dimension to the length of the gray shaft (Figure 6.10).

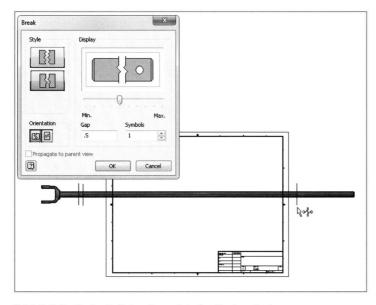

FIGURE 6.9 Defining the points for the break view

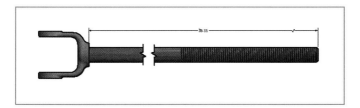

FIGURE 6.10 Adding a dimension to the break view

The dimension will show the true length of the visible portion of the bar.

Slice View

A *slice* view is different from a section view in that only the geometry of the cut portion is visible. It can be used for something like a rotated section or to modify another view at cutting planes using a sketch.

1. Verify that the 2013 Essentials project file is active, and then open c06-09.idw from the Drawings\Chapter 6 folder.

2. Start the Slice tool from the Modify panel on the Place Views tab.

3. Click the isometric view as the view to be modified, and select the check box in the dialog box for Slice The Whole Part.

    Once the target view is selected, you need to select a sketch that will be used to cut the view.

4. Select the vertical lines in the side view of the part, and click OK. Figure 6.11 shows the result.

FIGURE 6.11 A slice view can show how the shape of a component is structured.

This updates the isometric view to show the portions of the part where the lines pass through it.

Custom View

The ViewCube in parts and assemblies aligns with the standard view options for placing drawing views. At times, these views just can't properly describe the geometry. For these times, you can create a *custom* view.

1. Verify that the 2013 Essentials project file is active. If you created a new template in Chapter 5, "Customizing Styles and Templates," use Essentials 1.idw to start a new drawing. If not, start a new drawing using the ANSI (mm).idw template available on the Metric tab.

2. Start the Base View tool (in the Create panel of the Place Views tab or on the marking menu).

3. When the Drawing View dialog box opens, use the Browse tool to locate the files. Find c06-01.ipn in the Assemblies\Chapter 6 folder.

    The .ipn extension is for Inventor presentation files. A presentation file is used to create an exploded view of an assembly.

4. Once the exploded assembly is selected, its preview appears. Below the list of standard view orientations is the Change View Orientation button. Click it.

   Certification
   Objective

   This opens the presentation file in a specialized Custom View tab with file-viewing tools.

5. Position the exploded view similar to Figure 6.12.

6. Once the view is positioned, click the Finish Custom View tool at the end of the Custom View tab.

7. The preview updates to the view you established. Set the scale so the exploded view of the assembly fits across the page, as shown in Figure 6.13. Make it a shaded view.

8. Click the location for the view on the drawing sheet, and then right-click and select OK to finish the command and place the view.

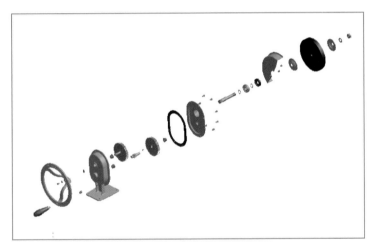

FIGURE 6.12 Positioning the parts for a drawing view

The presentation file can also create animated assembly instructions, and you will do that in Chapter 13, "Creating Images and Animations from Your Design Data."

T I P   A nice but less commonly used capability is to copy drawing views on the same page. This can be handy if you need to duplicate views to distribute detailing or want to save time setting up an alternative presentation of the views.

Using Advanced Drawing Annotation Tools

In Chapter 2, I covered the elements of dimensioning that will make up the bulk of typical detailing. Many other tools are necessary to make a complete drawing. The next group of exercises will focus on these tools.

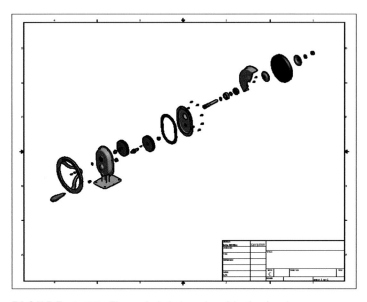

FIGURE 6.13 The exploded view placed in the drawing

Automated Text

In Chapter 5, you augmented a title block to include the values of the part's Mass properties. You can add these values to annotation text as well.

1. Verify that the 2013 Essentials project file is active, and then open c06-10.idw from the Drawings\Chapter 6 folder.

2. Just above the title block is a line of text. Double-click the text to open the Format Text dialog box.

3. In the Format Text dialog box, change the Type drop-down to Physical Properties—Model and the Property drop-down to Mass.

4. Click the Add Text Parameters button to add the model property to the text-editing window below.

5. Click OK to update the text.

The value of the mass appears, but it will likely be listed as N/A because the part's Mass properties are out-of-date. To update them, you need to open the part file and update it.

6. Right-click one of the drawing views on the drawing sheet or in the browser, and select Open from the context menu.

7. When the part file opens, right-click the Part icon at the top of the Browser, and select iProperties. This opens the c06-09.ipt iProperties dialog box.

8. Click the Physical tab, and click the Update button.

9. Return to the drawing to see the results.

THE APPROXIMATE WEIGHT OF THE FINISHED PART IS 0.237 lbmass

10. Close both the drawing and the part file without saving the changes.

Leader Text

You can use the Text tool to place text in a rectangular space on the drawing. The Leader Text tool automatically places the text at the end of a leader.

1. Verify that the 2013 Essentials project file is active, and then open c06-11.idw from the Drawings\Chapter 6 folder.

2. You can find the Leader Text tool on the marking menu. You can also find it in the Text panel on the Annotate tab. Start the Leader Text tool.

Leader
Text

3. Click the top machined surface of the part, and then click a second point to locate where the text will be generated.
   You can continue to click points to string the leader through.

4. Right-click and click Continue from the context menu to open the Format Text dialog box.

5. Type in anything you like, and then click OK to generate the text. See Figure 6.14 for an example.

TIP **Notice that you can create multiple rows of text in the leader.**

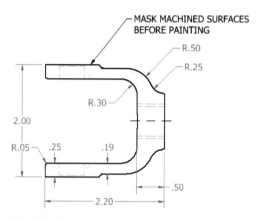

FIGURE 6.14 Text attached to the leader

Special Symbols

Many types of symbols are needed for drawings. Inventor includes a wide selection of symbols and the ability to define your own. These symbols can be placed directly or attached to a leader.

1. Verify that the 2013 Essentials project file is active, and then open c06-12.idw from the Drawings\Chapter 6 folder.

2. In the Symbols panel of the Annotate tab, locate and start the Surface tool.

3. Click the top surface that the Leader Text is attached to.

4. Right-click and select Continue from the context menu to open the Surface Texture dialog box (Figure 6.15).

5. Make any changes in the dialog box, and click OK to place the symbol.

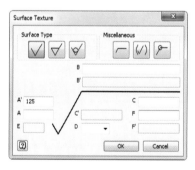

FIGURE 6.15 The Surface Texture dialog box is very thorough.

6. The Symbol tool will continue to work. Click the face on the far right, drag a leader away from the part, and click a second point for the leader.

7. Right-click and click Continue to launch the dialog box again, and place a Surface symbol on the leader. See Figure 6.16.

8. Right-click and choose Cancel, or press the Esc key to stop the Surface symbol tool.

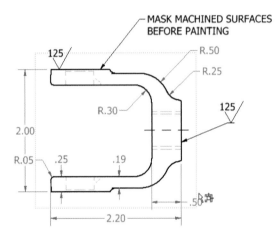

FIGURE 6.16 Symbols can be attached to leaders as well.

Although this exercise covers only one type of symbol, nearly all of them work in the same way.

Balloon

The exploded presentation view is a great way to display the physical makeup of an assembly, but you also need to be able to document what the components are. Calling out the parts using *balloons* and generating a list of the parts can help create a more complete document.

1. Verify that the 2013 Essentials project file is active, and then open c06-13.idw from the Drawings\Chapter 6 folder.

Certification Objective

2. The Balloon tool is on the marking menu (in an open part of the Design window) and on the Annotate tab in the Table panel. Start the tool.

Balloon

3. The Balloon tool opens in component-selection mode. Click the wheel on the left end.

This launches the BOM Properties dialog box to define the source of the balloon information based on the bill of materials (BOM) structure.

4. Leave the BOM View value as Structured, and click OK.

   As you place a balloon, it moves smoothly but snaps at 15-degree increments. Before the position of the leader is set, you can override the snap. Hold the Ctrl key as you position the balloon, and it can be placed at any angle.

5. Click a second point to set the length of the leader.

6. Right-click and click Continue to place the balloon. Then, press the Esc key to end the Balloon tool. See Figure 6.17.

7. Click the leader of the balloon, and the grip points will highlight. Click the point at the tip of the arrow, and drag it to the housing to the right.

8. When the edge of the part highlights, release the mouse to attach the balloon to the other part.

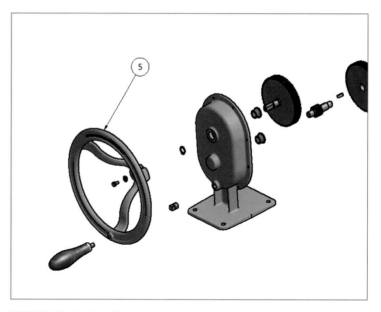

FIGURE 6.17 Placing a balloon in the drawing view

The balloon number will update to correspond with the value gathered from the bill of materials contained in the assembly.

Auto Balloon

The Balloon tool is effective, but a faster alternative is Auto Balloon.

1. Verify that the 2013 Essentials project file is active, and then open c06-14.idw from the Drawings\Chapter 6 folder.

2. The Auto Balloon tool is on the Annotate tab in the Table panel. You will have to expand the Balloon tool flyout to access it. Start the Auto Balloon tool.

   This tool works a little differently; it opens a dialog box (see Figure 6.18) immediately. The dialog box has a lot of options for determining how to prioritize what information is entered into the balloons.

3. Click the drawing view if the Select View Set button is depressed (see Figure 6.18). This activates the Add or Remove Components tool.

4. Drag a window around several components toward the right end of the exploded view. See Figure 6.19.

FIGURE 6.18 The Auto Balloon dialog box

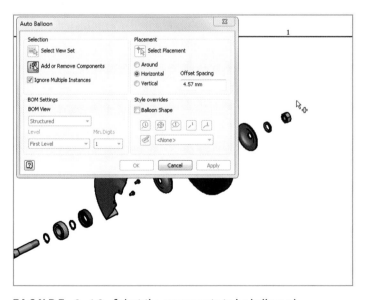

FIGURE 6.19 Select the components to be ballooned.

If you right-click a balloon, you can attach other balloons by clicking another part.

5. In the Placement group of the dialog box, set Offset Spacing to 10. Then click the Select Placement icon, and move your cursor over the screen to see a preview of the balloons.

6. Click to place the balloons on the screen, and then click OK.

Creating a Parts List

Certification Objective

The *parts list* is a way of documenting the bill of materials of the assembly. You do not have to present the entire BOM and are not limited to items contained in it.

1. Verify that the 2013 Essentials project file is active, and then open c06-15.idw from the Drawings\Chapter 6 folder.

Parts
List

2. Find the Table panel on the Annotate tab, and start the Parts List tool.

3. The Parts List dialog box that opens (Figure 6.20) has a number of options similar to those in the Auto Balloon dialog box.

FIGURE 6.20 The Parts List dialog box has many options.

4. For simplicity, you can leave the options alone. Click the drawing view, and then click OK.

5. Place the parts list in the drawing.

6. Drag the left edge of the column to resize the various columns (Figure 6.21).

7. Moving over portions of the text presents a cursor that allows you to click and drag the location of the parts list.

8. Locate the parts list above the title block and against the right border of the drawing. The parts list will snap into place. See Figure 6.22.

ITEM	QTY	PART NU
1	1	Grinder Base
2	1	Grinder Face
3	1	Spur Gear2
4	1	Gasket
5	1	Flywheel Handle

FIGURE 6.21 Drag the separation lines to resize the parts list's columns.

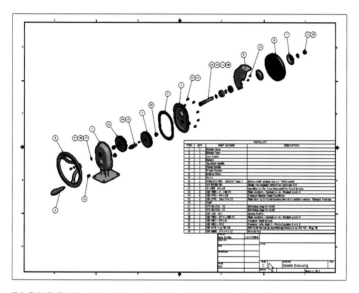

FIGURE 6.22 The parts list added to the drawing

 TIP If the parts list is based on the structure of the BOM, that structure can be represented in the parts list as well. The parts list would show a parent component and the tree of its child components.

The initial size of the columns and other properties of the parts list are controlled by the style and standards. This is another element that will likely be set once within a company and left alone.

Editing Dimension Values

Certification
Objective

When you set the dimension styles, the purpose was to make as many dimensions as possible appropriate for your needs. That doesn't mean you won't need to make changes to some.

1. Verify that the 2013 Essentials project file is active, and then open c06-16.idw from the Drawings\Chapter 6 folder.

2. Double-click the 66.00 dimension. This opens the Edit Dimension dialog box.

3. Choose the Precision And Tolerance tab, and set Tolerance Method to Symmetric and the Upper value to 0.01, as shown in Figure 6.23.

4. Click OK to update the dimension, as shown in Figure 6.24.

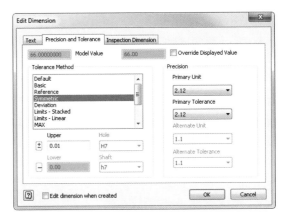

FIGURE 6.23 The Edit Dimension dialog box

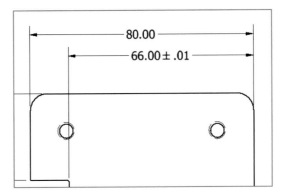

FIGURE 6.24 The dimension updated with the tolerance

There are also other ways to dimension the part.

Placing Ordinate Dimensions and Automated Centerlines

Ordinate dimensions (like baseline dimensions) can be placed either as individuals or as a set. The Ordinate Set option keeps the members styled as a set and also automatically does some spacing of the dimensions.

1. Verify that the 2013 Essentials project file is active, and then open c06-17.idw from the Drawings\Chapter 6 folder.

2. Right-click within the boundary of the view, and select Automated Centerlines from the context menu.
   The Automated Centerlines dialog box (Figure 6.25) allows you to select whether to include cylindrical extrusions or faces that have a radius, and you can also choose whether to select objects with an axis that is normal or parallel to the view.

FIGURE 6.25 The Automated Centerlines dialog box

3. Keep the defaults, and click OK to place the centerlines.

4. Locate the Ordinate dimension tool in the Dimension panel on the Annotate tab, expand it, and select the Ordinate Set tool.

5. In the status bar, notice the prompt to click an origin point. Click the center mark of the large hole in the lower-right corner of the part.

6. Then, click the vertical edges of the part shown in red in Figure 6.26 and the center marks highlighted with green dots.

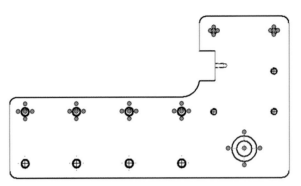

FIGURE 6.26 Selecting elements of the drawing view on which to place ordinate dimensions

7. When you've selected all the elements, right-click and select Continue from the context menu.

8. Drag the location of the dimensions above the model geometry, and click to place the dimensions.

9. Right-click and select Create to complete the dimensions. Figure 6.27 shows the result.

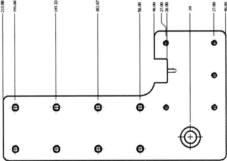

FIGURE 6.27 The ordinate dimensions placed in the drawing

You can drag ordinate dimensions to change their spacing, and grips allow you to change the location of jogs in the dimension line.

Ordinate dimensions can be placed in more than one direction on the view.

Hole Tables

Certification
Objective

When you have a large number of holes, it is more efficient to use a hole table than to dimension the location of each one.

1. Verify that the 2013 Essentials project file is active, and then open c06-18.idw from the Drawings\Chapter 6 folder.

2. The Hole Table flyout is located in the Table panel on the Annotate tab. Expand the list, and click the Hole View option.

3. Select the drawing view of the part.

4. Set the origin location to be the center of the large hole in the lower right of the part.

5. When the preview of the table appears, drag it near the top of the title block against the left edge of the border, and click to place it.

   Closer inspection of the drawing will show that it has labeled each of the holes with an alphabetical hole class and then a number that counts how many holes of that size are in the view. The table shows those hole names along with the dimensional information for placing the holes, as well as the description of each hole. To simplify the table, you can group the descriptions together.

6. Double-click the hole table to open the Edit Hole Table: View Type dialog box. Select the Combine Notes radio button (Figure 6.28) in the Row Merge Options group.

7. Click OK to update the table. Figure 6.29 shows the result.

FIGURE 6.28 Merging hole descriptions can make the table less confusing.

HOLE TABLE			
HOLE	XDIM	YDIM	DESCRIPTION
A1	-196.00	-12.50	
A2	-149.33	-12.50	
A3	-102.67	-12.50	
A4	-56.00	-12.50	Ø6.60 THRU
A5	-196.00	32.50	
A6	-149.33	32.50	
A7	-102.67	32.50	
A8	-56.00	32.50	
B1	-27.00	32.00	
B2	27.00	32.00	
B3	27.00	67.00	M6x1 - 6H
B4	-27.00	102.00	
B5	27.00	102.00	
C1	.00	.00	Ø13.49 THRU ⌴ Ø22.23

FIGURE 6.29 The revised hole table showing the combined notes

There are still more types of annotation that are more specialized, and I will cover them along with the design techniques they document in later chapters.

THE ESSENTIALS AND BEYOND

For many CAD users, creating detail drawings isn't just a final step; it is the most critical. Many companies use these drawings not only to produce their products but as the legal documentation of their products. It takes a strong tool set to produce them accurately and efficiently. Inventor has these tools.

ADDITIONAL EXERCISES

▶ Combine your learning from Chapter 5 and Chapter 6 to develop an appropriate Parts List style.

▶ Work with the Text tool to see how easy it is to build stand-alone notes.

▶ Explore and compare the behavior of the Ordinate and Ordinate Set tools.

▶ Test the various Tolerance and Fit callouts available in the Edit Dimension tool.

CHAPTER 7

Advanced Part Modeling Features

Now that you've developed the basic skills for part modeling and sketching, it is a good time to let you know that you've already learned some of the most challenging concepts in Inventor. Advanced features in Autodesk® Inventor® 2013 are as easy to define and often easier to construct than the basic features.

In this chapter, you will create or complete a handful of parts that include complex contours or large numbers of features. As you move beyond basic, prismatic parts, you need to work with more developed sketched and placed features. You will find that the basics learned with primitives, extrusions, and revolved features will apply directly to the dialog boxes and workflows of these more advanced tools.

In addition, by completing this chapter and Chapter 3, "Learning the Essentials of Part Modeling," you will have used the majority of solid modeling tools available in Inventor.

▶ **Projecting sketches and lofting**

▶ **Building a hole pattern**

▶ **Exploring advanced efficiency features**

Projecting Sketches and Lofting

For your first challenge, you will construct a spoke for a hand wheel that will change shape as it passes along a path. To do this, you will use a loft feature, but first you need to develop some of the elements required to form the feature.

Projecting a 3D Sketch

You can use a 3D sketch to create complex geometry for sketched features. It can be created with a stand-alone tool or as an option to a 2D sketch that defines part of it.

To define the second profile, you need to build a plane on a curved face:

1. Make certain that the 2013 Essentials project file is active, and then open the c07-01.ipt file from the Parts\Chapter 7 folder.

2. Double-click Loft Sketch 2 in the browser to edit the sketch.

3. Locate the Project Geometry tool in the Draw panel of the Sketch tab. Expand it, and start the Project to 3D Sketch tool.

4. In the dialog box, select the Project check box (Figure 7.1), click the curved face, and click OK.

   This will create a new 3D sketch, applying the circle to the cylindrical face.

5. Click Finish Sketch in the Ribbon or use the marking menu.

A 3D sketch can also be created on its own without the use of a 2D sketch.

FIGURE 7.1 Projecting a 2D sketch onto a 3D face

Creating the 3D sketch has given you a second profile for your loft. Now you need a path for the two profiles to transition along.

Defining a Loft Path Between Points

To define a loft along a path, you must create a minimum of three sketches. You need a beginning profile and an end profile, as well as another sketch to define the path along which the shapes will transition from one to the other.

The type of path you will be using must be built between existing points, ensuring alignment of the profiles you will loft between.

1. Make certain that the 2013 Essentials project file is active, and then open c07-02.ipt from the Parts\Chapter 7 folder.

2. Create a new sketch by selecting the XZ plane in the Origin folder of the browser and clicking the Create Sketch icon that appears in the Design window.

Project Geometry

Now, you will need to use the Project Geometry tool, which is on the Sketch tab in the Drawing panel and on the marking menu.

3. Use the Project Geometry tool to project the center point of the 3D sketch and then to project the point at the center of the red sketch on the wheel into the current sketch.

4. Create a sketch using the parameters and constraints shown in Figure 7.2.

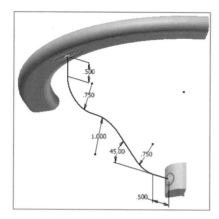

FIGURE 7.2 The sketch of the loft path

The sketch consists of three arcs and three lines. Remember that you can create a tangent arc by clicking the endpoint of a line and dragging the arc.

5. Be sure to finish the sketch. When you are modeling, you cannot save a file while a sketch is active.

Now you have the pieces in place, so you can create your loft features.

Creating Loft Features

Loft features are created using the Loft tool, which offers several options. For example, you can define whether sketches or model faces are used for profiles (sections) and whether the paths will connect to the geometry. You will now use this tool to create the wheel spoke.

Certification Objective

1. Make certain that the 2013 Essentials project file is active, and then open c07-03.ipt from the Parts\Chapter 7 folder.

2. Start the Loft tool from the Create panel of the 3D Model tab.

      The dialog box that opens (Figure 7.3) displays two primary areas for sections and rails or centerlines. As you click geometry, the selected parts will appear in these windows.

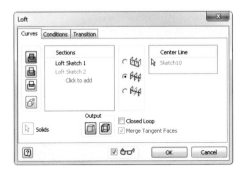

FIGURE 7.3 The Loft dialog box showing the Center Line path option

      When you start the tool, it will look to select sections first.

3. For the first profile, click the red sketch on the rim of the wheel segment. It is labeled as Loft Sketch 1 in the browser, or you can select it from the Design window.

4. Click the 3D sketch (Loft Sketch 2) for the second profile.

 5. Between the Sections and Rails groups are the Rail options. Set the option for Center Line. This will also change the Loft tab displayed in the dialog box to a Curves tab.

6. On the Curves tab, click where it says *Select a sketch* in the Center Line group. Select the sketched path.

      As you've seen with other sketched features in Inventor, a preview will appear when all of the elements needed to define it are satisfied.

7. Once a preview appears like Figure 7.4, click OK to create the feature.

 8. Drag the end-of-part marker to the bottom of the browser, or right-click and select the Move EOP To End option from the Context menu. The model will generate the pattern of the full solid model. See Figure 7.5.

The loft has created a connection between the hub and rim of the wheel. The minimum of two profiles were used, but you can add as many profiles as you need to accurately develop the shape you want. One loft option makes it possible to control the area of each section for more advanced designs.

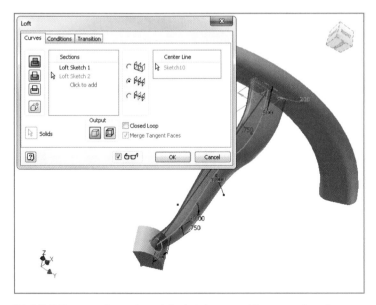

FIGURE 7.4 A preview of the loft feature will appear after the sketches are selected.

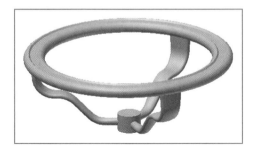

FIGURE 7.5 Patterning the entire solid allows you to edit a smaller feature set.

Using the Engineer's Notebook

To assist others who might need to edit your models in the future, it can be helpful to document the process or thinking behind a design. Inventor has a tool called the Engineer's Notebook that enables you to capture comments or images of your design at any time and keep them with the component for future reference.

1. Make certain that the 2013 Essentials project file is active, and then open c07-04.ipt from the Parts\Chapter 7 folder.

2. Expand the Notes element in the browser, and double-click Note 1 to open the notebook (see Figure 7.6) included in this file.

3. Right-click in the open notebook and choose Comment from the marking menu, or select the Comment tool from the Notes panel of the Engineer's Notebook tab.

4. Draw the rectangle that forms the comment, and add text similar to Figure 7.7.

5. Pick View from the marking menu or from the Notes panel, and create a view of the Graphics window.

By right-clicking and using the context menu, it is possible to orbit, pan, or zoom the view to get a better image, as shown in Figure 7.7.

6. Right-click in an empty area of the notebook and select Freeze from the context menu to preserve this view at this point in the model development for future reference.

7. Select Finish Notebook from the Exit panel to return to editing the model.

FIGURE 7.6 The contents of the notebook in the file show its history.

A small, yellow icon appears on your model. You can double-click it to access the notebook, as well.

TIP It is possible to create more than one Engineer's Notebook in a single component for different reasons. Notes can be created at any time and can be renamed or removed, if needed.

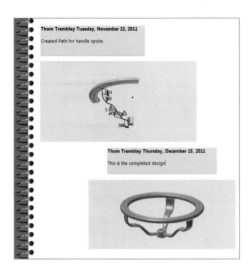

FIGURE 7.7 Be sure to capture a good description of changes in text and images.

Creating a Sweep

To create a sweep, you define a profile, a path, and, optionally, a *guide rail* or *guide surface*. For this exercise, you will use the optional guide rail, which will steer the profile as it sweeps along a path.

Certification
Objective

1. Make certain that the 2013 Essentials project file is active, and then open c07-05.ipt from the Parts\Chapter 7 folder.

2. Start the Sweep tool from the Create panel of the 3D Model tab.
   Since there is only one closed profile in the model, it will automatically be selected for the profile.

3. Select the dashed line as the path. This preview will appear like an extrusion.

4. Change Type from Path to Path & Guide Rail.
   When you change the option, an icon appears with a red arrow, which indicates that Inventor needs input from the user. You must satisfy all of these red arrows to complete the feature.

5. Select the curved line that is part of a 3D sketch as the guide rail.

6. Click OK to create the new feature, as shown in Figure 7.8.

The basic sweep uses a single profile, and the Path And Guide Surface option uses a model face to change how the profile transitions along the path.

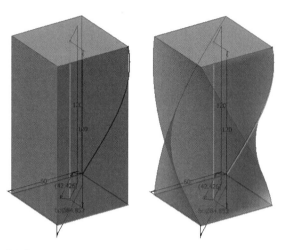

FIGURE 7.8 Using the guide rail allows the feature to be more complex.

The use of the 3D sketch's Helical curve allows the profile to twist around the path.

Creating a Shell

Certification
Objective

When parts are hollowed with a consistent wall thickness along a face, using the Shell tool is much more efficient than building the walls separately.

1. Make certain that the 2013 Essentials project file is active, and then open c07-06.ipt from the Parts\Chapter 7 folder.

2. Start the Shell tool from the Modify panel of the 3D Model tab.
   The Shell dialog box (Figure 7.9) opens, and you will be prompted to click the faces you want removed to create the hollowed shape and to name a thickness for the remaining walls.

3. Select the yellow face to be removed, and set the Thickness value to 5 mm by dragging the arrow or entering the value in the dialog box.

4. Expand the dialog box by clicking the icon on the bottom of the dialog box.

5. Click where the dialog box says *Click To Add*; then click the green face on the part, and set the value to 15 mm.

6. Click OK to create the shell feature, as shown in Figure 7.10.

You can also create a shell feature without a removed face to create a hollow part, and you can select multiple faces to open up the part as much as needed.

FIGURE 7.9 Expanding the Shell
dialog box gives access to additional
wall thicknesses.

FIGURE 7.10 The resulting shell
shows the difference in wall thickness.

Building a Hole Pattern

In Chapter 3, there were exercises for placing holes on a sketch that are concentric
to a curved face or edge. In this section, you'll explore an additional technique for
placing holes, namely, a *linear hole*, and then you will use the *rectangular pattern*
to replicate it. This is a common need on things such as mounting plates that need
to attach to several components.

Using a Linear Hole Placement

Linear hole placement requires selecting a face on which to place the hole. Once you've done that, you can simply click OK to generate the feature. You can also click one or two edges to locate the hole relative to them using parametric dimensions placed for you.

1. Make certain that the 2013 Essentials project file is active, and then open c07-07.ipt from the Parts\Chapter 7 folder.

2. Start the Hole tool by selecting it from the Modify panel on the 3D Model tab. It is also on the marking menu.

3. Set the hole class to Counterbore. Set the hole type to Clearance Hole, and then set the standard to be ANSI Metric M Profile. Set the fastener type to Socked Head Cap Screw, and set the size to M4 and the termination to Through All.

4. Click the top face of the plate near the notch.

5. Click the edges shown in Figure 7.11, and set their offset values to 14 mm.

6. Click OK to place the hole.

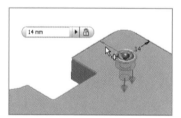

FIGURE 7.11 Placing the hole based on the edges of the part

The diameter and depth of the counterbore and the size of the hole were all determined based on standards for the fastener that will go through it. With the hole in place, you can replicate it using the Rectangular Pattern tool.

Creating a Rectangular Hole Pattern

Certification
Objective

Let's create a series of these holes. On this part, the holes will be in a pattern where the two directions are perpendicular, but the pattern can go in any two directions.

1. Make certain that the 2013 Essentials project file is active, and then open c07-08.ipt from the Parts\Chapter 7 folder.

2. Go to the top icon of the Pattern panel on the 3D Model tab, and click the Rectangular Pattern tool.

3. The dialog box requires you to select a feature, so click the counter-bore hole created in the previous exercise.

4. Click the button with the red arrow in the Direction 1 group, and then click the short edge from which the counterbore hole was positioned.

5. Set the number of instances to 3 and the distance between them to 34.

6. Click the icon for Direction 2, and select the longer edge from which the hole was positioned.

7. To set the proper direction of the new instances, click the Flip button.

8. Set the number of instances to 2 and the distance to 52 per Figure 7.12.

9. Click OK to place the pattern on the part.
   One of the occurrences in the pattern is not needed. It interferes with other features of the part.

10. Locate the Rectangular Pattern1 feature in the browser, and expand it.

11. Hover over the Occurrence icons until the hole that overlaps the notch in the part highlights, as shown in Figure 7.13.

12. Right-click and click Suppress on the context menu. The completed feature will look like Figure 7.14.

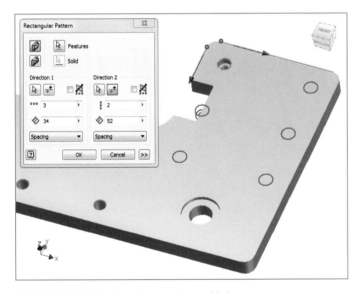

FIGURE 7.12 Creating a pattern of holes

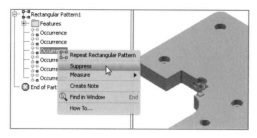

FIGURE 7.13 Individual instances in a pattern can be suppressed.

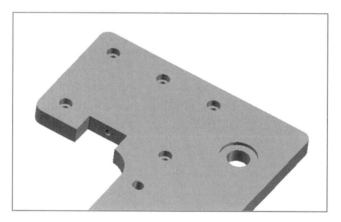

FIGURE 7.14 The hole pattern with one occurrence suppressed

It is possible to suppress any number of members and to change whether they're suppressed at any time.

Expanding on the Rectangular Pattern

The fundamentals of the Rectangular or even the Circular Pattern are much the same. As mentioned in the previous exercise, the directions of the pattern don't have to form a rectangle.

In this exercise, not only will you use one direction, but you will see that direction can be a curve.

> **1.** Make certain that the 2013 Essentials project file is active, and then open c07-09.ipt from the Parts\Chapter 7 folder.

2. Start the Rectangular Pattern tool from the Pattern panel of the 3D Model tab.

3. Select the Hole Boss and Hole 1 features from the browser.

4. In the Rectangular Pattern dialog box, make the direction one selection and pick the projected sketch that runs along the edge of the opening.

5. Set the number of instances to 6, and use the drop-down to change the spacing to Curve Length, as shown in Figure 7.15.

   The initial preview shows instances of the features in space. It is usually necessary to define the start point of the pattern to make sure it follows the curve properly.

6. Expand the dialog box, pick the selection icon for the Start of Direction 1, and then select the point at the center of the hole.

7. When the preview updates showing the instances following the sketch, click OK to create the new features.

   The new hole bosses and holes will be created, but there is one more change to make to the model before you continue.

8. Expand the Extrusion1 feature in the browser, and edit Sketch 1.

9. Change the 2.5-inch dimension to 4 inches, and finish the sketch. To see what the updated model will look like, take a look at Figure 7.16.

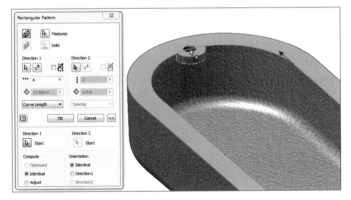

FIGURE 7.15 Be sure to select the sketch for your path and the hole center for a start point

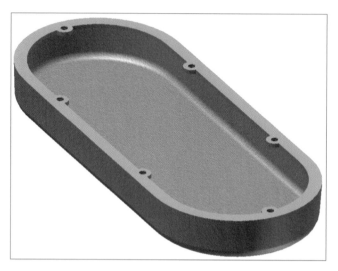

FIGURE 7.16 Updating the part updates the sketch and therefore the pattern.

Using a sketch as the path for a pattern opens a lot of options for creating complex patterns easily.

Exploring Advanced Efficiency Features

Among Inventor's advanced features are editing tools that enable you to increase efficiency by replicating features, editing features after their creation, and even replacing a portion of an existing feature to give it a new contour.

In the next group of exercises, you will convert a simple block into a complex bottle using simple tools. You will see how a good workflow allows the complex problem to be broken down into a few simple steps.

Combining Fillet Types

Placing individual types of fillets is an easy process, but sometimes combining types of fillets can allow you to build. That's how you'll begin to add some nice shape to your bottle.

    1. Make certain that the 2013 Essentials project file is active, and then open c07-10.ipt from the Parts\Chapter 7 folder.

    2. Orbit the part so that you can see the right and top face.
       For this exercise, you will use the names of the ViewCube faces to identify the faces on the part and the edges where they intersect.

3. Start the Fillet tool, which is in the Modify panel of the Model tab or in the marking menu.

   When the tool starts, a dialog box and a mini-toolbar will appear. For the exercise, you will focus on the mini-toolbar.

 4. In the mini-toolbar, use the drop-down to change the fillet type to Full Round Fillet.

5. Click the Right, Top, and Left faces of the part. As soon as you click the third, the preview of the fillet will appear.

 6. Move back toward the mini-toolbar, and it will expand. Click the Apply button (green plus sign) to add the fillet, and continue selecting other options.

 7. On the mini-toolbar, change the fillet type to Edge Fillet.

8. Return to the Home view, click the edges shown in Figure 7.17, and set the radius to 3. Do not click OK or Apply.

FIGURE 7.17 Place a constant radius on the model edges.

 9. In the mini-toolbar, change the type of edge fillet to Add Variable Set, and click the edge shown in Figure 7.18.

10. Click near the middle of the curved edge to place an additional radius reference point.

11. Set the radius to 6 and the edge position to .5, as shown in Figure 7.18, but do not click OK or Apply yet.

    The final step, adding a setback, can also be done with the mini-toolbar, but you will use the dialog box instead.

The standard fillet type will resolve the edges with tangency. There is also an option with edge fillets to apply G2 continuity for smooth transitions.

12. Click the Setbacks tab in the dialog box. Click Vertex 1 at the intersection of the Bottom, Front, and Left faces. Set the Setback values for the three edges to 8.

13. Pick Click To Add below Vertex 1 to add another vertex at the intersection of the Bottom, Front, and Right faces. Set their values to 8 as well, as shown in the dialog box in Figure 7.19, and then click OK.

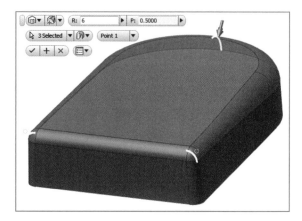

FIGURE 7.18 Add a point along the edge to add another radius.

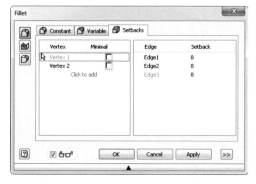

FIGURE 7.19 The Setbacks tab of the Fillet dialog box doesn't offer more options than the mini-toolbar but can present them better.

Setbacks add a softer corner and are popular on items such as consumer products. By building the constant and variable radius features at once, setbacks can be included.

Adding a Draft Angle

When creating an extrusion, sweep, and a few other features, you can include a draft angle, referred to in their dialog boxes as a Taper value. Sometimes it is best to add draft after the fact. With the bottle, for example, you need to include draft on a face that couldn't be added at the time the main feature was created. For that purpose, you can use the Draft tool.

1. Make certain that the 2013 Essentials project file is active, and then open c07-11.ipt from the Parts\Chapter 7 folder.

2. Be sure that you orbit the part to see the green face on the bottom.

3. Start the Face Draft tool on the Modify panel of the Model tab.
   The Face Draft dialog box will prompt you with an icon to select the pull direction. The pull direction is a face or plane around which the draft angle pivots.

4. Select the yellow face on the part to use it for the pull direction.

5. Select the side face of the recessed extrusion near the yellow face, and set the angle to 15 degrees, as shown in Figure 7.20.
   Once selected, the preview of the face raft should show the bottom edge of the face moving into the open space. If it doesn't, select the Flip Pull Direction icon to correct it.

6. Click OK to add the draft to the part.

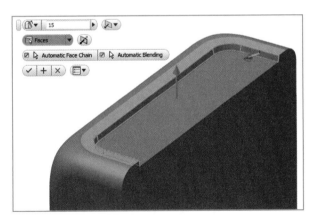

FIGURE 7.20 Like some other features, draft provides an arrow you can drag to change its angle.

You can also place the draft based on a plane or a sketch. The plane can even be set away from the faces to be drafted.

Replacing One Face with Another

Certification
Objective

In cases where a complex shape makes up only a portion of a part, you can replace a planar face with a contoured one. To continue building the bottle, you'll use that technique to build a curved face that couldn't readily be created with the original feature.

1. Make certain that the 2013 Essentials project file is active, and then open c07-12.ipt from the Parts\Chapter 7 folder.

   At the bottom of the feature tree is a surface feature created by extruding an open profile. It is also possible to revolve and loft a surface by selecting the option.

2. Click the surface that passes through the part.

3. Select the Edit Sketch tool from the screen.

4. The model will reposition so the view is normal to the sketch.

5. Press F7 (or right-click and select Slice Graphics) to remove the solid model down to the sketch, as shown in Figure 7.21.

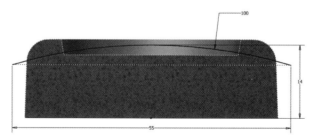

FIGURE 7.21 The Slice Graphics function can expose sketches within the model.

You will notice that there is sketch geometry that makes up the perimeter of the portion of the part that the sketch intersects. This geometry cannot be created using the normal Project Geometry tool. It has to be created using Project Cut Edges.

6. Use Finish Sketch to return to the 3D model.

7. Expand the Surface panel on the 3D Model tab, and select the Replace Face tool.

8. Click the large recessed face on the front of the part.

9. In the dialog box, click the New Faces selection button, and then click the surface.

10. Click OK to update the model.

11. In the browser, right-click the ExtrusionSrf1 feature, and then deselect Visibility.

Another tool that can be used for this is Sculpt, which you'll use in Chapter 9, "Creating Plastic Parts." Replace Face is more basic and works well for substituting geometry.

Mirroring

Pattern tools reduce the tedium of creating repeated features. Creating a mirror feature can save work when there is symmetry in the part.

Certification Objective

1. Make certain that the 2013 Essentials project file is active, and then open c07-13.ipt from the Parts\Chapter 7 folder.
   The Mirror tool is located in the Pattern panel of the 3D Model tab.

2. Start the Mirror tool.

3. In the Mirror dialog box, select the Mirror A Solid option.
   Choosing this option will limit your selection option to clicking the mirror plane.

4. Click the XY Plane under the Origin folder, and click OK to create the full body shown in Figure 7.22.

FIGURE 7.22 Use the Mirror tool instead of modeling both sides.

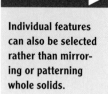

This part could've been modeled as a quarter of the body, mirroring it around the YZ plane and then the XY plane. Once the symmetrical portions are mirrored, you can add unique features to the different areas of the part.

Creating an Offset Work Plane

Work planes are needed when faces don't exist on which to create sketches or to use to develop other features. These can be developed by selecting edges, points, or other planes.

1. Make certain that the 2013 Essentials project file is active, and then open c07-14.ipt from the Parts\Chapter 7 folder.

   The Work Plane tool is available on the marking menu. More options are available in the Work Features panel on the 3D Model tab (Figure 7.23).

FIGURE 7.23 Filters make creating work planes easier.

2. Expand the Origin folder in the browser.

3. Start the Offset from Plane option from the Work Planes drop-down, and click the XZ plane in the browser.

4. Click and drag the new plane until its value reads 80 or type in the value.

 5. Click the OK icon to create the plane.

Similar filters can also be found for work axis and work point features.

The work plane you created will be used to build a sketch. It is also possible to start a new sketch, click a face or plane, and drag as you did in this exercise and simultaneously create a new plane and sketch on it.

Using a Fillet to Close a Gap

If you look closely at the model you are about to open, you will see a gap between the cylinder on the top and the main body. In these situations, applying an edge fillet is not possible because the faces do not intersect. In these cases, a face fillet will bridge a fillet across gaps.

1. Make certain that the 2013 Essentials project file is active, and then open c07-15.ipt from the Parts\Chapter 7 folder.

2. Start the Fillet tool from the marking menu or the Ribbon.

 3. Change the option in the mini-toolbar (or dialog box) to the Face Fillet option.

4. Click the cylindrical face of the Top Boss feature and one of the large curved faces just below it.

With the default options, you need to click only two faces.

5. Selecting the faces causes the tool to preview a fillet that can solve between the faces.

6. Set the radius to 3, and click OK from the mini-toolbar (Figure 7.24) to generate the fillet.

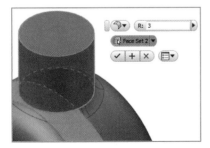

FIGURE 7.24 Creating a fillet between faces can close a gap.

You can use this option to close large gaps or problematic edges.

Adding a Coil

Threaded hole features do not generate physical threads, but if you need threads or to represent a coil of any type, there is a specific feature for it. In this case, you'll use a coil to create threads for the mouth of the bottle.

1. Make certain that the 2013 Essentials project file is active, and then open c07-16.ipt from the Parts\Chapter 7 folder.

 2. Start the Coil tool from the Create panel on the 3D Model tab.

3. The profile will be automatically selected. Click the y-axis from the Origin folder for the axis of the coil. It might be necessary to use the direction button in the dialog box to make sure the coil is going toward the body of the part.

    This will open the Coil dialog box. This dialog box has three tabs for selecting geometry, specifying the coil, and controlling the ends of the coil.

4. Switch to the Coil Size tab and set the type to Revolution And Height, set the Height value to 7, and set the Revolution count to 2, as shown in Figure 7.25.

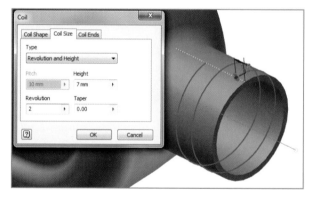

FIGURE 7.25 Using the Coil tool to generate threads

5. Click OK to generate the coil feature.

This feature can be created using pretty much any profile as long as it doesn't intersect itself. This makes it flexible for many things.

Using Open Profiles

All of the sketched features you've created have used closed profiles to generate solid features and open sketches to create a surface feature. An open profile is capable of creating a solid when it can intersect with existing solid geometry. You'll take advantage of that to create a stop on the neck of the bottle.

1. Make certain that the 2013 Essentials project file is active, and then open c07-17.ipt from the Parts\Chapter 7 folder.

2. Set your view to look at the front of the ViewCube.

3. Start the Revolve tool from the Create panel on the 3D Model tab.

4. Click the open profile that is in the sketch and move the mouse toward the solid until it shows a preview of the area the sketch that will be projected toward the body of the part. See Figure 7.26.

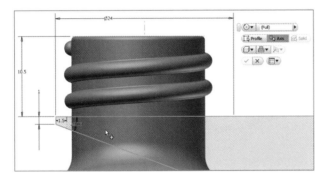

FIGURE 7.26 Previewing a projected profile

5. Click the mouse when the correct area is highlighted to use this profile.

6. Click the y-axis in the browser for the axis or the centerline in the sketch.

7. Once selected, the preview will change color. Click OK to create the complete feature. Figure 7.27 shows your finished part.

Open profiles work well for dealing with parts that include draft on faces. It is difficult to align a close profile with these faces.

FIGURE 7.27 The finished part showing the revolved lip

View Representations in a Part

In Chapter 8, "Advanced Assembly and Engineering Tools," you will use view representations to save different configurations for the way an assembly appears. In a part file, you can do the same thing to quickly switch between preselected color or texture options.

1. Make certain that the 2013 Essentials project file is active, and then open c07-18.ipt from the Parts\Chapter 7 folder.

2. Expand the View options in the browser, and confirm that a view representation called Aqua is the active version.

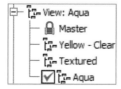

3. Switch between the other saved options to see the options.

If your workflow allows a part to have different colors without having a different part number, view representations can give you the option of placing the part in an assembly using different representations to show different colors.

THE ESSENTIALS AND BEYOND

The advanced filleting tools in conjunction with lofting and sweeping tools open many doors for exploration in part modeling. Patterns and mirrored parts and features are great ways of simplifying and speeding up the creation of parts. Used in combination, these tools make the possibilities endless.

ADDITIONAL EXERCISES

▶ Look at some of your current designs for opportunities to use mirroring and patterning.

▶ Try using fillets instead of sketched curves.

▶ Shell features can be used to create thin parts by removing all but a few faces. Consider whether you have parts on which you could use this technique, and give it a try.

▶ Using work features is a critical skill to master. Practice using them and combining them.

CHAPTER 8

Advanced Assembly and Engineering Tools

Most of the tools needed for the day-to-day work of building assemblies were covered in Chapter 4, "Putting Things in Place with Assemblies." In this chapter, you will use a few more tools to build your assembly.

Being able to navigate in the assembly and maintain your system's performance is also important, and large assemblies can put you at risk of losing data if your system runs out of memory. Autodesk® Inventor® software representations can solve these problems, and you will see how to control your assembly with them.

Much of the focus of this chapter will be on tools that allow you to work more efficiently by creating geometry in the context of the assembly. These tools also offer a unique opportunity by incorporating engineering calculators into them. You will see how to create common machine components, but you will also see how to make sure they will do the job.

▶ **Controlling the assembly environment**

▶ **Using design accelerators**

▶ **Working with additional assembly tools**

Controlling the Assembly Environment

An often-overlooked aspect of working with assemblies is controlling both the look-and-feel of the assembly's view representation and, most importantly, its system demands. Both of these elements are easy to control but are not automatic. There are three kinds of *representation* in Inventor that enable this enhanced level of control: view, level of detail, and positional. In this section, you'll try the first two.

Creating View Representations

View representations allow the user to control the visibility and appearance of components in the assembly. Creating several view representations allows a user or multiple users to see the assembly in whatever way they want.

1. Make certain that the 2013 Essentials project file is active, and then open the c08-01.iam file from the Assemblies\Chapter 8 folder.

2. Locate the Representations folder in the Browser. Expand the folder, and expand the View group.

3. Right-click the Default view representation, and select Copy from the context menu.

   This creates a new view representation named Default1 at the bottom of the list.

4. Click the new view representation twice slowly, and rename it Natural Colors.

5. Double-click the browser icon of the new view representation to activate it.

6. Right-click the Natural Colors representation, and select Remove Appearance Overrides from the context menu.

7. Select c08–01:1 in the Browser, right-click, and then uncheck Visibility in the context menu. Figure 8.1 shows the results.

8. Double-click the icon next to Default to set it as the active view representation, and then switch the view representation back.

FIGURE 8.1 View representations can save variations of color and visibility.

The updates to the assembly affect only the visibility and colors of the components in the assembly. Another form of representation has a stronger effect.

Creating Level-of-Detail Representations

Level-of-detail representations will not just remove a component from visibility; they will remove it from memory. If you need to work on large assemblies with limited system resources, the level-of-detail (LOD) representation is an invaluable tool.

Certification
Objective

1. Make certain that the 2013 Essentials project file is active, and then open the c08-02.iam file from the Assemblies\Chapter 8 folder.

2. Locate the Representations folder in the Browser. Expand the folder, and expand the Level of Detail group.
   On the right end of the status bar, two set of numbers report the number of components displayed in the Design window and the number of documents being accessed by Inventor.

3. Double-click the icon next to the Gears representation.
   This reduces the number of displayed component occurrences and the number of open documents by removing all the components other than the shafts and gears that drive the mechanism.

4. Activate the All Content Center Suppressed representation and see the effect.

5. Right-click the All Content Center Suppressed LOD, and select Copy from the context menu.

6. Rename the new LOD Primary Components.

7. Double-click the browser icon of the Primary Components LOD to activate it.

8. In the Design window, click the wooden handle.

9. Right-click, expand the Selection options in the context menu, and select Component Size to be the selection filter for the assembly.
   This opens a dialog box showing the volume of the selected part.

10. Change the selection value to 205, and click the check mark to select the components smaller than 205 millimeters.

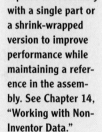

> **A substitute LOD can replace the assembly with a single part or a shrink-wrapped version to improve performance while maintaining a reference in the assembly. See Chapter 14, "Working with Non-Inventor Data."**

11. Right-click and select Suppress from the context menu. The effects are shown in Figure 8.2.

FIGURE 8.2 LOD representations can remove components from memory.

You must save your work to preserve the changes made to an LOD representation before switching to another one.

Certification
Objective

POSITIONAL REPRESENTATIONS

Positional representations can override whether a constraint is enabled or suppressed. They can also use different offset values for a constraint. For example, a positional representation named Open can change the value of a Mate constraint from 0 mm to 10 mm, which will cause a gap to appear between parts. These representations can be used to create overlays in a drawing to show an alternative position or be used in an animation (see Chapter 13, "Creating Images and Animation from Your Design Data") to reduce the number of steps needed to create the animation.

Using Design Accelerators

Certification
Objective

Inventor has a number of specialized tools that build common machine components. Design accelerators come in two basic classes: calculators that work out what geometry you will need (such as weld sizes) and generators that can simply create components or use their built-in calculation tools to help you select the appropriate components such as bolts or the size of a shaft.

It helps to build something when learning advanced tools. In this chapter, you will be working on the design of a hand-driven metal grinder.

Using the Bearing Generator

Many bearing standards are global, so this tool allows you to choose from a number of different components with the same basic geometry.

1. Make certain that the 2013 Essentials project file is active, and then open the c08-03.iam file from the Assemblies\Chapter 8 folder.

2. Click the Bearing tool in the Power Transmission panel of the Design tab. The Bearing Generator dialog box has two tabs. The Design tab displays a bearing (selected based on the size specified), and the Calculation tab allows you to verify that the bearing will hold up to the demands of its job. All generators have at least these two tabs.

3. Set the dimensional values for the dialog box to be the same as Figure 8.3, and click the Update button to build a list of bearings that meet the physical specs.

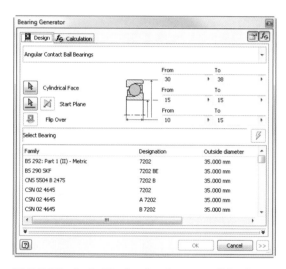

FIGURE 8.3 The Bearing Generator dialog box can quickly generate a list of bearings based on size.

4. Click the down arrow near the top of the dialog box where it says *Angular Contact Ball Bearings* to open a selection dialog box.

5. In the dialog box, use the drop-down menu in the upper right to change the category of bearing to Cylindrical Roller Bearings, and then select the Cylindrical Roller Bearings (Figure 8.4) option in the upper left so that it creates a list of all of the bearings that fit.

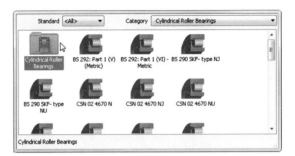

FIGURE 8.4 You can change the category of bearing that you place in the assembly.

By default components placed by design accelerators might not relocate if geometry they are aligned with is changed in the assembly. You change this option by right-clicking the component in the browser, going to the component flyout, and selecting Automatic Solve.

▶

Certification Objective

c8-10:1

6. Scroll down the list, select the SKF NJ bearing, and click OK to create the geometry. You will have to hit OK in the File Naming dialog box as well.

7. Click to place the bearing off to the side of the cover.

Using the Calculation tab, you can also test the bearing by checking how the bearing meets life span, loading, and speed ratings to be sure that it will hold up.

Using the Adaptivity Feature Within the Assembly

Adaptivity is not a function of a design accelerator, but it is an advanced capability unique to Inventor. It allows part features to be modified by assembly constraints in certain conditions.

It is possible to identify an adaptive item in the Browser by locating the red-and-blue icon next to the part or feature or sketch.

1. Make certain that the 2013 Essentials project file is active, and then open the c08-04.iam file from the Assemblies\Chapter 8 folder.

2. Switch the Ribbon to the Inspect tab, and start the Distance option of the Measure tool from the Measure panel, press the M key, or select Measure Distance from the marking menu.

Distance

3. Click the inside edge of the large hole in the tan part.
   The hole in the part will measure 38 mm. The bearing in the assembly has an outer diameter of 35 mm. Adaptivity will allow the diameter of the hole to be changed to fit the bearing.

4. Start the Constrain tool in the Position panel of the Assemble tab.

Constrain

5. Hover over the exterior cylindrical face of the bearing.

6. When the axis of the bearing highlights, right-click and click Select Other from the context menu.

7. Click the first Face option from the list that highlights the outer diameter of the bearing, as shown in Figure 8.5.

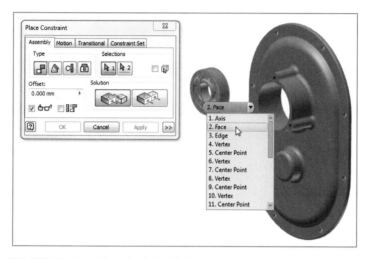

FIGURE 8.5 Use the Select Other option to click additional selection options for placing assembly constraints.

8. Select the inside face of the large hole in the front cover, and click OK.

9. Click the front face on the ViewCube so you can see that the diameter of the hole is matched to the diameter of the bearing.

Adaptivity is a great tool for sizing one component based on another, but using it on too many features or parts can slow your performance. It is best used to establish a size, and then it can be disabled to lock that size in.

Using the Shaft Generator

It is a simple process to create a shaft by revolving a half section or creating a series of extrusions stacked end to end. The Shaft Generator design accelerator constructs the shaft through a dialog box and lays out the segments of the shaft.

1. Make certain that the 2013 Essentials project file is active, and then open the c08-05.iam file from the Assemblies\Chapter 8 folder.

2. Hold down your Ctrl key and click the Shaft tool in the Power Transmission panel of the Design tab.

Shaft

Holding the Ctrl key is necessary only if you want to start a design accelerator with the default options.

When the dialog box (Figure 8.6) opens, shaft sections appear in the large window in the dialog box. Each segment row has columns for the left and right end treatments, the section shape, and any special additions. The size and shape are listed to the right as well.

FIGURE 8.6 The Shaft Generator dialog box showing the default shaft segments

   3. Select the top segment. At the far right, click the Delete icon, and then click Yes in the dialog box that appears.

4. Delete all but the segment listed as Cylinder 55 × 100.

5. Double-click the description of this segment.

6. Change the D (diameter) and L (length) values to 10 mm by clicking the current values in the Cylinder dialog box that appears.

7. Click OK to update the size of the segment.

   8. Locate and select the Insert Cylinder icon just above the updated section.

9. This will insert a duplicate segment below the original. Edit the size of the new segment to D = 12, L = 3, and click OK.

10. Click the Insert Cylinder tool again, and create a new segment with D = 25 and L = 12 values.

11. Create another new cylinder, setting D = 12 and L = 16.

12. Create the fifth segment to be the same as the first: D = 10 and L = 10.

13. In the bottom segment, click the end treatment icon on the right, and click the Chamfer option in the drop-down.

14. When the small dialog box appears, set the Distance value of the chamfer to .3, and click the OK check mark.

15. Add a .3 chamfer to the left end of the first shaft segment. The Shaft Component Generator dialog box now should look like Figure 8.7.

FIGURE 8.7 Adding chamfers to the ends of the shaft

16. Click OK to generate the shaft and OK again to accept the names of the new files.

17. Place the new shaft into the assembly by clicking a location in the Design window.

The Shaft Generator is great for hubs, too. A drop-down in the Sections group allows you to add internal geometry from either end.

At this point, you've developed the geometry that fits in the assembly and that you think will handle the loads. Part of every generator in the Power Transmission tools can also calculate whether the geometry is up to the task at hand. Now, you should see whether this shaft will work.

Calculating and Graphing Shaft Characteristics

For this exercise, you will continue using the shaft you are in the middle of defining. The Shaft Component Generator's Calculation and Graph tabs have nothing to offer until you've defined geometry on the Design tab. In turn, the Graph tab cannot supply information until the calculation is done.

1. Right-click the shaft you placed in the assembly, and select Edit Using Design Accelerator from the context menu.

2. Click the Calculation tab to open the options available. See Figure 8.8.

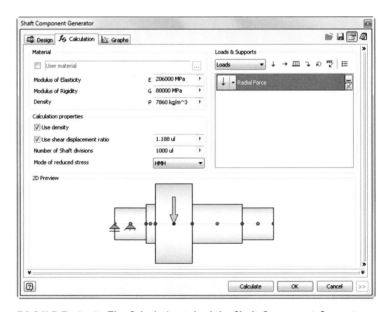

FIGURE 8.8 The Calculation tab of the Shaft Component Generator

The Calculation tab is where you can override and experiment with the material of the shaft and set up how it is supported and loaded.

3. Click the check box next to the Material callout. This will open a Material Types dialog box.

4. Select Steel, and click OK to make it the active material.

5. In the 2D Preview, click the Free support icon at the far-left end of the preview, and drag it to the middle of the rightmost section. See Figure 8.9.

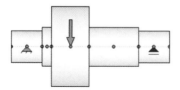

FIGURE 8.9 The 2D preview shows how the shaft is loaded and supported.

6. Click the Load icon, and the Loads & Supports area will change to show the type of load that is being used.

7. Double-click the Load icon to open the dialog box, showing the properties of the load.

8. Change the force to 200 N, and click OK to close the dialog box.

9. Click the Calculate button, and then click the Graphs tab.

> Calculate

10. The Shear Force graph shows the results.

11. Click the Deflection graph in the Graph Selection area to see that the deflection of this shaft is 0.811 micron (Figure 8.10).

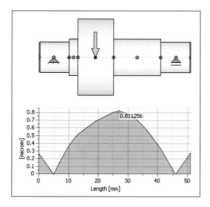

FIGURE 8.10 The Graphs tab displays the results of the calculations done on the shaft.

12. Click OK to close the dialog box.

The Calculation tab varies based on the tool, but the information that you can enter or extract is precisely the information you would need in order to use engineering criteria to evaluate the components you're creating. You won't use the Calculation tools for all the tools I will review, but keep in mind that they are available.

Using the Spur Gear Generator

The Design tab of this dialog box is broken down into three major sections. The top is the Common area where you specify the tooth characteristics of the gear. The middle is where you set the requirements for size of the gears that support the teeth, and the bottom is an information area that will offer feedback on the design and calculations.

Spur
Gear

1. Make certain that the 2013 Essentials project file is active, and then open the c08-06.iam file from the Assemblies\Chapter 8 folder.

2. Hold the Ctrl key, and click the Spur Gear tool on the Power Transmission panel of the Design tab.

    The Spur Gears Component Generator dialog box opens and displays the default values (Figure 8.11).

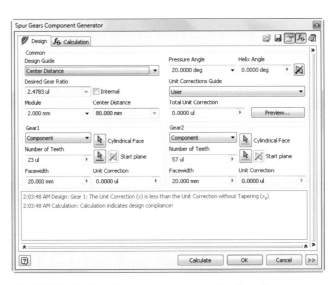

FIGURE 8.11 The default values for the Gear Generator

3. Beginning in the upper left, change Design Guide to Number Of Teeth.

4. Set the Desired Gear Ratio value to 4.0000 ul from the drop-down.

5. Set Module to .8 mm and Center Distance to 50.000 mm.

    Now you need to make changes to the specifications for the individual gears.

6. Use the drop-down to change Gear1 and Gear2 from being component to features. The Component option will build a new part in the assembly that can be fitted to the shaft. The Feature option will modify the shaft, and the No Model setting is used as a reference to develop the gear that mates to its values.

7. Set the Facewidth value for both gears to 13 mm, and your dialog box should now look like Figure 8.12.

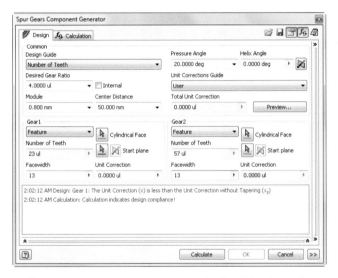

FIGURE 8.12 The updated values in the dialog box prior to selecting the geometry

8. Under Gear1, click the selection icon for the cylindrical face, and then select the cylindrical face on the largest-diameter segment of the gray shaft.

9. Click the selection icon for the start plane of Gear1, and click the flat face on the same shaft segment.

10. Start the Cylindrical Face selection for Gear2, and click the large, cylindrical face on the green shaft.

11. For the start plane of Gear2, click the large face on the same segment of the green gear.

12. Click the Calculate tool to update the geometry preview.

     The lower section will indicate a design failure. You might have to select the double arrows to expand the dialog box to see the error in the lower section.

13. Click OK to generate the gear features on the existing shafts, and click Accept to generate the gears (Figure 8.13) even with the incomplete information.

FIGURE 8.13 The updated shafts with the gear features added

Anyone who has created gears in 2D or 3D by developing a tooth profile and patterning it will appreciate the power of the Gear Generator. Along with spur gears, there are design accelerators for worm and bevel gears.

Using the Key Connection Generator

Keys are used to align components but typically need material removed from at least two parts to be able to place the key. In this case, a Gear component has been added to the assembly centered on the shaft.

1. Make certain that the 2013 Essentials project file is active, and then open the c08-07.iam file from the Assemblies\Chapter 8 folder.

2. Double-click the large gear.

The gear will be solid with no opening for the shaft or accommodation for reducing the weight of the gear. Since the part is a solid model, you can use any solid modeling tool to change it, but in this example, you will create the shaft accommodation using the Key Connection Generator.

3. Click the Return icon on the Ribbon or click Finish Edit in the marking menu to return to the assembly.

4. Switch the Ribbon to the Design tab, hold the Ctrl key, and click the Key Connection tool in the Power Transmission panel.

The Parallel Key Connection Generator dialog box (Figure 8.14) has some similarities to the Bearing Generator dialog box. The process is straightforward. Select the type of key you want, and then tell it what kind of groove you want for the shaft and hub or if you want one.

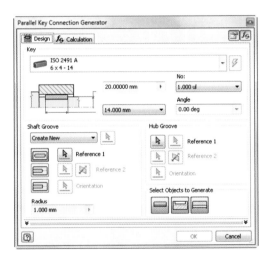

FIGURE 8.14 The Parallel Key Connection Generator dialog box default values

5. At the top of the dialog box, the current key standard is displayed. Click the down arrow at the end of this space to open the selection dialog box.

6. At the top of the dialog box showing the available key types, click the Standard drop-down, and select ANSI to limit the options.

7. Click the Rectangular or Square Parallel Keys type.

8. Just below the key selection is the value of the shaft diameter. Change this from 20.000 mm to 12 mm.

   The next several steps define where the key is positioned and which parts of the assembly will be changed. Selecting the correct geometry is critical. Figure 8.15 shows where to click to properly locate the key.

In the dialog box, you can choose whether to generate the hub groove, shaft groove, or key.

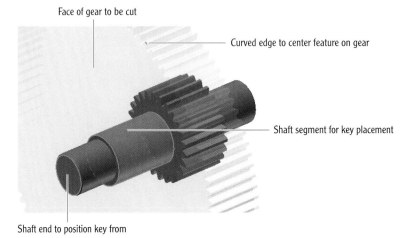

Face of gear to be cut

Curved edge to center feature on gear

Shaft segment for key placement

Shaft end to position key from

FIGURE 8.15 Select only the faces and edge (step 4) indicated.

9. Click the selection icon for Shaft Groove Reference 1, and then click the shaft segment between the gear and the smaller end segment.

10. The selection for Reference 2 automatically activates. Click the end of the shaft.

11. Reference 1 for the hub groove is now active. Click the face on the large gear.

12. For Reference 2 of the hub groove, you need to select a curved edge. Carefully select the tip of one of the gear teeth. If you click a side edge, simply click the selection icon and try again. You can zoom in to make it easier.

    Once the fourth selection is made, a preview of the key and grooves appears. You want to make one change to the shaft groove before you proceed.

13. In the Shaft Groove group, click the groove with one Rounded End option.

14. There are two arrows on the ends of the preview of the key. Drag the left arrow roughly 11 mm from the origin to the right.

15. Then drag the right until the key changes to its 11.113 mm standard size, as shown in Figure 8.16.

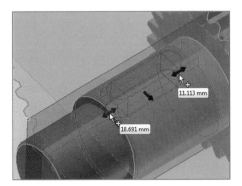

FIGURE 8.16 The position and length of the new key can be changed by dragging icons in the preview.

16. Click OK to generate the geometry, and click OK again to approve the creation of the new files. See Figure 8.17 for the finished key component.

FIGURE 8.17 The finished key with the groove built into the shaft

Both the shaft and the hub will have new features added to them, and a new part will be created for the key. Editing the key's size will update all of the related features.

Working with Additional Assembly Tools

Constraints and design accelerators are great, but they're not the only capabilities Inventor provides for an assembly. Some of these tools are not even unique to the assembly environment, but they simplify the process of defining your products.

For the next exercises, you will use a gear housing that is machined from a casting. As you start, you have a casting for only one half of the part, and it may need reinforcement.

Mirroring Components

When modeling parts, taking advantage of symmetry or the ability to pattern a feature can save work and improve quality. The same is true when it comes to whole components that are opposites of one another. For these components, you can mirror the entire part. In the context of an assembly, you can mirror several parts at once and generate the new files for them.

1. Make certain that the 2013 Essentials project file is active, and then open the c08-08.iam file from the Assemblies\Chapter 8 folder.

[icon] Mirror

2. Start the Mirror Components tool from the Component panel of the Assemble tab.

3. The Mirror Components: Status dialog box will be prepared to select components to be mirrored. Click the housing and the two bearings in the Design window.

4. Then click the selection icon for the mirror plane, and click the XY plane from the Origin folder of the assembly.

   A preview of the mirrored components will appear, and they will be color-coded based on the action to be taken. Green components will create a new file that is a mirror image of the original. Yellow components are new instances of existing geometry. This is the default state for standard components such as bearings.

5. In the dialog box, click the yellow icon next to the first bearing.

The Ignore option in the Mirror Component dialog box can be used to skip parts accidentally clicked with the window or crossing selection.

6. The icon turns gray, which indicates that the components will be ignored by the tool.

7. Your screen should look like Figure 8.18. Click Next.

   The Mirror Components: File Names dialog box allows you to specify the filename and path for the components that are to be created.

You can also apply a new naming scheme to all the new components by adding a prefix or suffix to them.

8. Click OK to create the new files and produce the new components in the assembly, as shown in Figure 8.19.

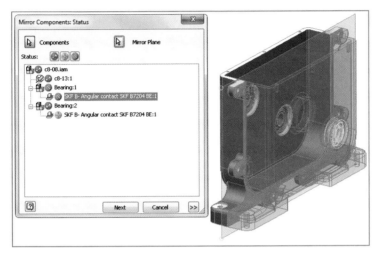

FIGURE 8.18 A preview of the mirrored and instanced components

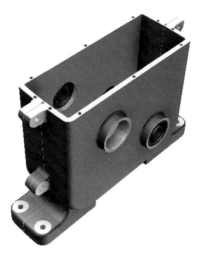

FIGURE 8.19 The mirrored housing in the assembly

Now that you have the mirrored part, you can make changes to the new part to make it uniquely different from the original, if need be. It is important to use the side that is least likely to need uniquely different features as the basis for the mirrored component to keep from having to remove geometry not needed in the copy.

Deriving Components

The Derive Component tool is one of the most powerful in Inventor. It can be used for everything including linking the value of one parametric part into another, creating a mirror of a component, and, as in this exercise, using one part as the basis for another. The concept is to begin with the geometry, sketches, dimensions, or faces from one part and use them in another to ensure consistency.

1. Make certain that the 2013 Essentials project file is active. From the Quick Access toolbar, start the New File tool, and create a new file using the Standard (mm).ipt template.

2. Switch the Ribbon to the Manage tab, and start the Derive Component tool in the Insert panel.
   Derive is also on the 3D Model tab in the Create panel.

3. Use the Open dialog box to select c08-12.ipt from the Parts\ Chapter 8 folder.

4. Click Open to go to the next step.
   The Derived Part dialog box (Figure 8.20) will allow you to choose what form your linked data will take. You can even choose to import only faces, sketches, or parameters. Once you've selected how you want to bring it in, you can change its scale or even mirror it around a plane.

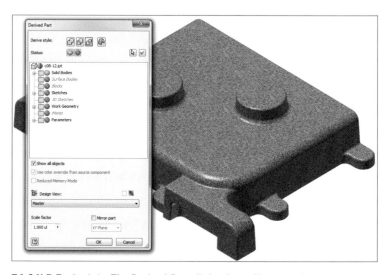

FIGURE 8.20 The Derived Part dialog box offers complete control over the reuse of existing data.

5. Leave the default options, and click OK to completely link the c08–12 part into the new part.

You can now add features to the linked geometry to make a machining of this cast part. Now, let's open an existing machining and the original casting.

You can use Derive Component to create a single or multi-body part out of an assembly even as a mirrored or scaled geometry.

6. Open the c08–13.ipt file from the Parts\Chapter 8 folder.

Take a moment to review the holes, extrusions, and other features that have removed material from the original casting.

7. Open c08–12.ipt from the Parts\Chapter 8 folder.

8. In the Browser, locate the end-of-part marker. Click and drag it to the bottom of the list of features.

After the part updates, you will see that a rib feature has been added to the model.

9. Click the File tab below the Design window to display your new part. It will not show the change to the casting yet.

10. Click the local update icon in the Quick Access toolbar to bring the changes into your new part.

11. Switch your Design window to the c08–13.ipt file and update it. See Figure 8.21.

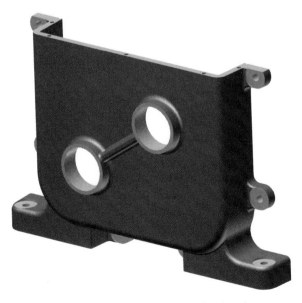

FIGURE 8.21 The updated part showing changes made to the casting file

TIP   If you use a single casting as the base for multiple components, the Derive Component tool will save you countless hours of updating machined parts. As mentioned in the introduction, this tool has many more uses and warrants further exploration.

Constraining and Animating Assembly Motion

 In Chapter 4, I covered the majority of Assembly constraint options that hold components in place. The Motion constraint can bind the movement of one component to another.

1. Make certain that the 2013 Essentials project file is active, and then open the c08-09.iam file from the Assemblies\Chapter 8 folder.

2. In the Position panel of the Assemble tab, click the Constrain tool.

3. In the Place Constraint dialog box, click the Motion tab.

 4. Set the Ratio value to 4, and then set the Solution value to Reverse. This tool is click-sensitive. The ratio is applied from the first selection to the second selection.

5. For the first selection, click a flat face or the end of the brass gear.

6. For the second selection, click the end of the gray shaft below it, as shown in Figure 8.22.

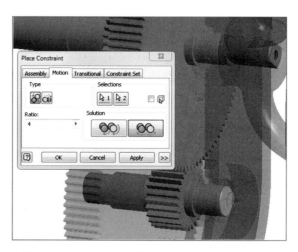

FIGURE 8.22 Selecting the components that will move together

7. Click OK to create the constraint.

8. Click and drag either the wheel or the wooden handle to see the mechanism work. Note that the last shaft turns at a final ratio of 16:1 over the handle.

9. Locate the Drive This Constraint item in the Browser under the New Shaft part.
   Even though this constraint is suppressed, it can still be used to animate the assembly.

10. Right-click the constraint, and select the Drive Constraint tool from the context menu.

11. In the Drive Constraint dialog box, set the End value for the range of motion to 360.

12. Expand the dialog box, and set the value for Repetitions to 3 as shown in Figure 8.23.

You can save the animation of a Driven constraint to an AVI or WMV file by clicking the Record button in the dialog box.

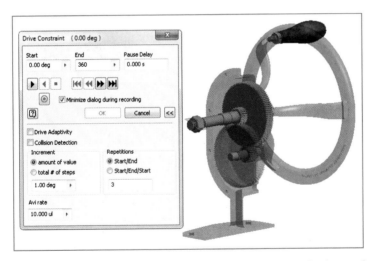

FIGURE 8.23 Driving a constraint can show how a mechanism works

13. Click the Forward button to see how the mechanism works.

Applying motion constraints to an assembly helps you verify that your design works the way you need it to work. Being able to animate this action can help you show others what you intend the design to do.

THE ESSENTIALS AND BEYOND

Representations can add control over how your assemblies look, but leveraging the LOD tools can improve the performance of your computer, save you time waiting for changes to update, and save you money by keeping your hardware effective longer.

Design accelerators add a whole new dimension to CAD, adding engineering tools to common component tools. Integrating these calculations may help you improve the efficiency of your products.

ADDITIONAL EXERCISES

▶ Practice using LOD representations in your day-to-day work. These tools provide many different advantages, including speeding up the drawing view by using them as the basis for view creation.

▶ Try using the Cam, Spring, and other tools available on the expanded Power Transmission panel.

▶ Explore different shaft loads and loading options to see how the shaft reacts.

▶ Look at the Calculation tab on all the design accelerators as you try them to see what information you will need to supply to allow the calculators to work.

CHAPTER 9

Creating Plastic Parts

A combination of specialized plastic part features and common solid modeling tools is needed to make any plastic part. Because certain types of geometry are common in plastic parts, automated tools were created to build this geometry quickly and easily. Taking features such as lips, bosses, and grills and then building wizards around them can save you hours on even a simple part.

In this chapter, you will learn about many of these specialized tools in the context of a workflow that will show a different approach to building components. There are many options that I will not cover, but you will see that these tools give you all the options you need in a way that is nearly self-explanatory.

▶ **Developing specialized features for plastic components**

▶ **Creating an assembly using a multibody solid to maintain a consistent shape**

Developing Specialized Features for Plastic Components

Plastic part features are specialized tools that create geometry that you could create using traditional features, but it would take many features to create geometry that is common in these types of parts. Figure 9.1 illustrates the model you'll complete in this chapter's exercises. To create and refine its shapes, you'll use many of the most common sculpting tools available in the Autodesk® Inventor® 2013 software.

FIGURE 9.1 The completed model from this chapter

Sculpting a Plastic Surface

In Chapter 7, "Advanced Part Modeling Features," you used the Replace Face tool to substitute a contoured face. The Sculpt tool could be considered an evolution of Replace Face, but it is far more powerful. It will help you combine a series of overlapping surfaces into a solid model. You'll use it to create the initial shape of your model.

1. Verify that the 2013 Essentials project file is active, and then open the c09-01.ipt file from the Parts\Chapter 9 folder.

2. Start the Sculpt tool from the Surface panel of the 3D Model tab.

3. Select all the surfaces in the Design window, including the plane at the base of the part.

4. As the surfaces are selected, a preview of solid bodies appears. Once all the surfaces are selected, click OK.

The complete solid should look like Figure 9.2.

Selecting the direction arrows in the surfaces allows you to bias the direction of the volume that will be solidified if there is any ambiguity.

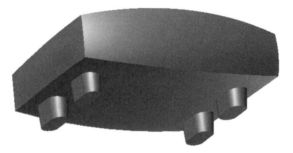

FIGURE 9.2 The solid formed by the surfaces

> **TIP** For creating components with a series of complex surfaces, it can be beneficial to create the surfaces as surfaces and sculpt them together instead of trying to construct a loft or other features.

Stitching Surfaces Together

You can combine surfaces to form a new surface, and then that surface can be used for other things. In this example, you'll use the Stitch tool to define a surface to be removed.

1. Verify that the 2013 Essentials project file is active, and then open c09-02.ipt from the Parts\Chapter 9 folder.

2. In the Browser, expand the Solid Bodies folder.

3. Right-click the icon next to Solid1, and uncheck Visibility to turn off the solid model.

4. Start the Stitch tool from the Surface panel of the 3D Model tab.

 Stitch

5. Click the extruded surface and the boundary patch; then click Apply to create the new surface.

6. Click Done to close the dialog box.

7. Make the solid visible again.

8. Start the Sculpt tool, and select the new surface.

9. Switch to the Remove option.
   A preview appears showing the portion to be removed in red (Figure 9.3).

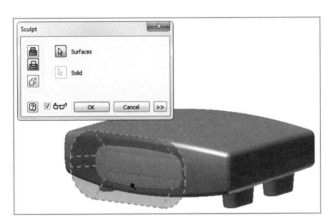

FIGURE 9.3 The Sculpt tool can also remove a portion of the solid using a surface.

10. Click OK to finish editing the part.

If you have a series of surfaces that meet at their edges, you can stitch them together into a solid. It will work, but cleaning up the edges of the surfaces is more difficult than allowing them to overlap and using the Sculpt tool.

TIP You can turn a single part into separate bodies representing and even defining the multiple parts of an assembly. By forming these various bodies using the same geometry, you can maintain the consistency of their shapes.

Splitting Bodies

You can generate almost any sketched feature as a new body by simply selecting the New solid option in the dialog box used to create that feature (i.e., extrude, revolve, loft). You can also use a plane or sketch geometry to split a face, remove a portion of a body, or even divide a solid model into separated bodies. To use the Split tool in this model, you will use a plane that will define the boundary between the two bodies:

1. Verify that the 2013 Essentials project file is active, and then open c09-03.ipt from the Parts\Chapter 9 folder.

2. Start the Split tool from the Modify panel of the 3D Model tab.

3. Click the Split Solid option in the dialog box, and click the visible work plane on the part or the Splitting Plane 1 feature in the Browser.

4. Click OK to break the part into two bodies.

5. In the Browser, expand the Solid Bodies folder to see the new solids that have been added.

6. Rename Solid2 to Front and Solid3 to Back by clicking the name twice slowly, as you would rename a file in Windows Explorer.

7. Click the solids individually from the Browser, and change their colors using the drop-down in the Quick Access toolbar. My choices are shown applied to the solid in Figure 9.4.

FIGURE 9.4 Once the solid is split into multiple bodies, you can treat them differently.

You can divide and subdivide a body as many times as needed to suit your needs. You can also use an extrusion to cut a part and then use the shared sketch to create a new body to form interlocking components.

Adding a Lip

A lip allows the plastic parts to remain aligned so the form remains consistent. You can use other sketched features, such as the Sweep tool, to make the geometry, but the Lip tool makes it easier to adjust things like draft and to recess edges.

1. Verify that the 2013 Essentials project file is active, and then open c09-04.ipt from the Parts\Chapter 9 folder.

2. Rotate the front body so that you can see the back side of it.

3. Start the Lip tool. It is on the 3D Model tab in the Plastic Part panel.

4. In the dialog box, make sure the option on the left is set to Lip, and then select the Pull Direction check box.
   This changes the dialog box options for constructing the lip feature.

5. Click the flat face around the perimeter of the part to set the Pull Direction option.

6. Once the direction is selected, click the inner edge of the perimeter for the path edges, as shown in Figure 9.5.

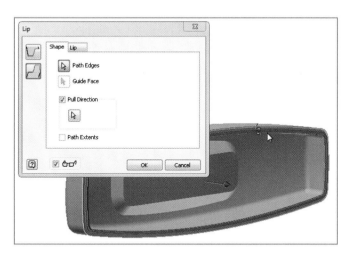

FIGURE 9.5 Selecting model edges to place a lip feature

7. Click the Lip tab in the dialog box, and review the options.

8. Click OK to place the feature using the default options. The result is shown in Figure 9.6.

FIGURE 9.6 The lip added to the front body

 TIP   You can also select the back part and place a lip using the Groove option. If a lip feature is placed in a part, its properties will be picked up by a groove in the same part.

Adding a Boss

Like the Lip tool, the Boss tool comes with opposite geometry sets. The head and thread sides of the boss are collections of complex geometry that would normally take several features to create.

The bosses you add in this exercise will allow you to assemble the halves of the back together using screws:

1. Verify that the 2013 Essentials project file is active, and then open the c09-05.ipt file from the Parts\Chapter 9 folder.

🔲 Boss

2. Start the Boss tool, which is in the Plastic Part panel on the 3D Model tab.

3. Make sure the Head option is active. It is the button in the upper-left corner of the screen.

4. Switch the Boss dialog box to the Head tab, and change the bottom two size options from 6.6 to 6 and from 7.54 to 7.

5. Expand the Draft Options section in the dialog box.

▶

You can also build ribs around the boss, and you can adjust their height, thickness, and angle.

6. Change the first two draft options from 2.5 to 2 degrees. Refer to Figure 9.7 for the dialog box values.

7. Click OK to generate the Head side bosses.

8. Expand the Solid Bodies folder, right-click the Back–Top body, and select Hide Others in the context menu.

This makes a new body visible and turns off the others. Now, you can add features to this body.

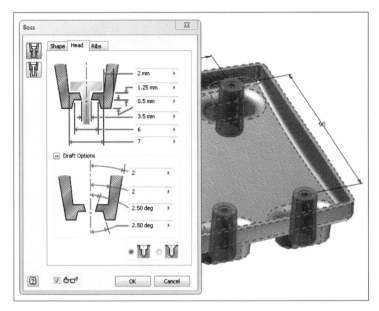

FIGURE 9.7 The Boss dialog box options can control many parametric values.

9. Scroll down in the Browser, and expand the Boss1 features.

10. Click and drag the sketch just above the Boss1 so that it is now shared.

11. Right-click the sketch, and make it visible so that it can be used to create a new feature.

12. Start the Boss tool, and switch it to the Thread option.

13. Set the value of the fillet to 3 (Figure 9.8), and click OK to place the new features.

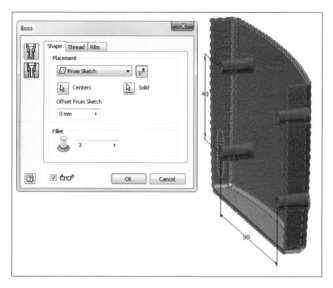

FIGURE 9.8 A boss can automatically add a fillet at the base for the new boss.

 TIP You can add stiffening ribs as part of the feature. You can also allow the hole to pass all the way through the component or be created only to a specified depth. By default the bottom of the hole will be set to and follow the contour of the existing face to reduce the amount of sink on the part.

Creating a Rest

A *rest* is like an extrude command that understands a wall thickness. It is meant to build a flat spot and can construct the internal and external faces at the same time.

1. Verify that the 2013 Essentials project file is active, and then open c09-06.ipt from the Parts\Chapter 9 folder.

2. Find and start the Rest tool in the Plastic Part panel of the 3D Model tab.

3. After the preview appears, click the direction button option in the dialog box on the left to put the hollow of the rest to the inside of the part.

4. Set the thickness to 2, and then click the More tab.

5. Set the values for Landing Taper and Clearance Taper to 2, as shown in Figure 9.9, and click OK to create the rest.

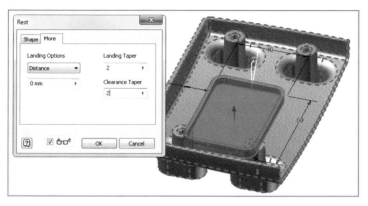

F I G U R E 9 . 9 The Rest tool will create a flat on the plastic part.

T I P You can also place a rest that intersects a face; in that case, it would simultaneously add and remove geometry in different directions.

The Rule Fillet Tool

The Fillet tool used in Chapter 3, "Learning the Essentials of Part Modeling," and Chapter 7 is powerful but can require a lot of input from the user. The Rule Fillet tool works based on relationships between the body of the part, its features, and its faces. You will use this tool to add fillets between a rest feature and the other features of a solid by specifying how the features blend, not the edges on which you want a fillet placed.

1. Verify that the 2013 Essentials project file is active, and then open c09-07.ipt from the Parts\Chapter 9 folder.

2. Start the Rule Fillet tool from the Plastic Part panel of the 3D Model tab. The primary options for the tool allow you to select either a feature or a face.

3. When the dialog box opens, click the rest feature, set Radius to 2, and leave the rule set to Against Part.
   The fillet preview appears anywhere the selected feature makes contact with the rest of the part.

4. Click below the first rule where it says *Click to add* to create a new rule.

5. Click the rest again, leave the radius at 1 mm, but change the rule to Free Edges. See Figure 9.10.

After the edges are selected, you can click a face or edges to be excluded from the rule fillet.

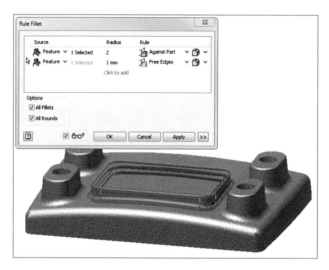

FIGURE 9.10 Adding a rule fillet between a feature and the part

This applies the second rule to any edge that is not tangent with another face or already slated to be affected by another rule.

6. Click OK to place the fillets based on these rules.

Expanding the Grill dialog box reveals a Flow Area calculation. This refers to the opening created by the grill, because grills are commonly used for ventilation.

> **TIP**   A rule fillet placed between features or between a feature and the body does not cause an error message if the conditions don't allow it to generate. Instead, it lies dormant until the conditions are restored and it can generate the feature again.

Adding a Grill

A grill is a feature that creates an opening in a part. It frequently has ribs across the opening. It can also include areas of solid obstruction called *islands*.

1. Verify that the 2013 Essentials project file is active, and then open c09-08.ipt from the Parts\Chapter 9 folder.

2. In the Design window, orbit the part so you can clearly see the sketch that is visible.

 Grill

3. In the Plastic Part panel of the 3D Model tab, start the Grill tool.
The Grill dialog box has five tabs with options for the sketch geometry you want to use for that subfeature.

4. The first tab sets the boundary. To define the boundary, click the circle in the sketch.

5. Click the Island tab, and for its profile, select the ellipse in the middle of the sketch. Set the thickness of the island to 1.

6. Select the Rib tab, and then click the six parallel lines in the sketch. Do not change any of the dialog box options.

7. Now, click the Spar tab, and pick the three lines that are perpendicular to the last lines. Set Spar Thickness to 2 and Top Offset to .5.

8. Compare your model to Figure 9.11.

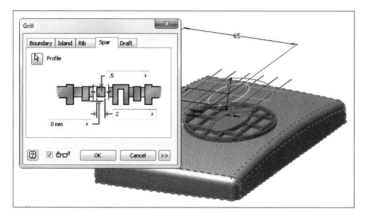

FIGURE 9.11 A single grill feature can replace several other features.

9. Expand the dialog box, and it will update to show what the open sectional area of the grill is, for ventilation considerations.

10. Click OK to generate the grill.

The grill is the most flexible feature for plastic parts. A wide selection of subfeatures allows you to pick and choose the elements you want to place.

Embossing or Engraving

The rules for creating an embossed or engraved feature, such as the product logo, are the same as if you were going to extrude a solid feature. You can use a closed sketch or text to define the feature's geometry.

1. Verify that the 2013 Essentials project file is active, and then open c09-09.ipt from the Parts\Chapter 9 folder.

2. Position the model so you can see the text in the sketch above the grill feature.

3. Start the Emboss tool from the Create panel of the 3D Model tab.

4. When the Emboss dialog box opens, click the text.

   The Emboss tool has three options. One will add geometry, one will remove, and the third will do both if the sketch plane intersects a face on the part.

5. Keep the Emboss From Face option active, and set the depth to .5.

6. Make sure the direction indicator is pointed toward the part (Figure 9.12). It might be necessary to change it. Then, click OK to create the feature.

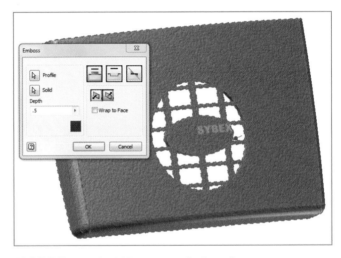

FIGURE 9.12 Adding text to the face of a part

TIP   The Emboss tool also has the ability to wrap a profile or text around a face so that the edges of the feature are normal to the surface they're wrapped on.

The Snap Fit Tool

The two options for the geometry for this tool are the hook and the loop. The features are created with a wizard, so all that is needed is a work point or a sketch point.

1. Verify that the 2013 Essentials project file is active, and then open c09-10.ipt from the Parts\Chapter 9 folder.

2. Orbit the model so that you can clearly see the open portion.

3. Start the Snap Fit tool from the Plastic Part panel on the 3D Model tab.

   A preview of two features appears, because there are two points in the sketch.

4. Open the Shape tab in the dialog box. Click the Flip Beam Direction button, and notice the change to the preview. If the original preview showed the hooks correctly positioned (protruding from the cover), click the Flip Beam Direction button again.

5. Return to the Hook tab. Notice the green arrows; the arrows indicate the direction of the hook. One of them points inward. Click that arrow until both hooks are pointing outward.

6. Switch to the Beam tab of the dialog box.

7. Match your dialog box values to those in Figure 9.13, and click OK to build the features.

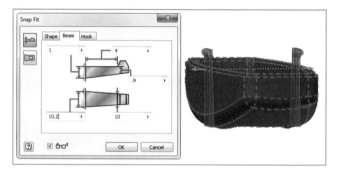

FIGURE 9.13 The Snap Fit tool creates complex geometry easily.

The loop side of a snap fit is placed in the same way. In general, all the plastic features follow a common process that makes them easy to learn.

 TIP   Another advantage of using a multibody solid is not having to reenter size information to create mating plastic features. The feature remembers the options changed in the active file.

Adding Ribs

The Rib tool is not a specialized plastic part feature, but it is common in plastic parts. Ribs are used in many types of parts, and the Rib tool can be used to create a rib or a web; you have the ability to control the width and depth.

1. Verify that the 2013 Essentials project file is active, and then open c09-11.ipt from the Parts\Chapter 9 folder.

2. Orbit the model so that you can clearly see the open portion.
   Sketched in the interior of the front part is a series of lines that will locate the ribs. Those lines do not extend to the edges of the part and

do not need to do so. The rib feature is capable of extending itself to fill in the geometry.

3. Start the Rib tool from the Create panel of the 3D Model tab.

4. Click the four lines that are partially obscured by the block in the middle of the part.

5. Set the thickness to 2. See Figure 9.14 for the preview of the feature.

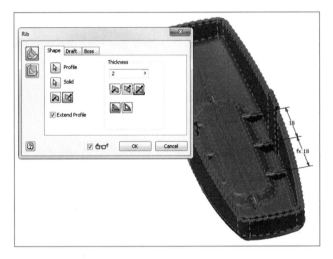

FIGURE 9.14 Rib features can extend beyond the sketch.

6. Click the OK button to create the ribs.

7. Orbit the part to see that the ribs did not penetrate through the front face of the part.

 TIP On parts with draft, it is difficult to properly create a sketch that doesn't risk creating a gap as it is created. Being able to create a rib with a sketch that doesn't meet the faces helps avoid problems when these parts change.

Adding Decals

It is possible to include geometry, text, or images in a sketch. A *decal* is a feature made from an image that is applied to flat or curved surfaces.

1. Verify that the 2013 Essentials project file is active, and then open c09-12.ipt from the Parts\Chapter 9 folder.

2. Create a new sketch on the XY plane.

3. Click the Image option in the Insert panel of the Sketch tab.

4. In the Open dialog box, navigate to the Parts\Chapter 9 folder, click the c09-01.png file, and then click Open.

5. Click a place in the Design window, and click to place the image. Hit Esc to finish placing.

   The size of an image can be controlled by a parametric dimension as well.

6. Place a dimension on the top of the image file, and set the value to 60. Hit the Esc key to finish the dimension tool.

7. You might have to use Zoom All to find the image in the sketch.

8. Drag the sketch so it is roughly centered on the recessed face of the part.

9. Finish editing the sketch.

10. On the 3D Model tab, expand the Create panel by clicking the arrow next to the panel name, and select the Decal tool.

11. Uncheck the Chain Faces option in the dialog box, and then click the image and the flat face behind it. See Figure 9.15.

12. Click OK to place the feature.

FIGURE 9.15 Selecting the face onto which to project the decal

TIP   This was a simple placement of a decal. The tool doesn't really shine until you start wrapping images around faces. The Decal tool is capable of that as well.

Creating an Assembly Using a Multibody Solid to Maintain a Consistent Shape

A multibody part simulates an assembly. It can be used to create an assembly. The assembly will be made up of components created from some or all of the bodies in the part. Any change to the bodies of the part will be reflected in the part created from it.

Converting Bodies to Components

Converting the bodies to parts requires the user to specify a new assembly, select the bodies, and name the new part files.

Make
Components

1. Verify that the 2013 Essentials project file is active, and then open c09-13.ipt from the Parts\Chapter 9 folder.

2. Switch the Ribbon to the Manage tab, and start the Make Components tool from the Layout panel.

3. When the Make Components dialog box opens, it is in selection mode. Click the three components from the Design window or from the Solid Bodies folder of the Browser.

4. In the dialog box, set the Target Assembly Name field to **New Assembly .iam** and the Target Assembly Location field to **C:\Inventor 2013 Essentials\Assemblies\Chapter 9**.

5. Compare your dialog box to Figure 9.16, and then click Next.

6. You don't need to change anything in the Make Components: Bodies dialog box. Click OK.

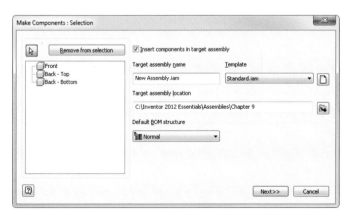

FIGURE 9.16 Set the name of the assembly and its path.

This creates the new files and then opens the new assembly with the related parts. The parts will be grounded into position, so applying assembly constraints isn't needed.

Draft Analysis

There are tools for several types of shape and geometry analysis in Inventor. You can analyze the curvature, the sections of the body, and even the continuity of the surfaces.

Draft analysis gives a basic understanding of how a part might separate from a mold. It gives you the opportunity to make changes early in the process.

1. Verify that the 2013 Essentials project file is active, and then open c09-14.ipt from the Parts\Chapter 9 folder.

2. Switch the Ribbon to the Inspect tab, and click the Draft tool on the Analysis panel.

Draft

3. Expand the Origin folder in the Browser, and select the XY plane for the direction to which the pull would be normal.

   In the Draft Analysis dialog box, you will find a scale showing the colors related to the draft angle. You can change the range of the positive and negative draft results and establish the range of allowed angles. Understanding the color related to the angle will help you interpret the results.

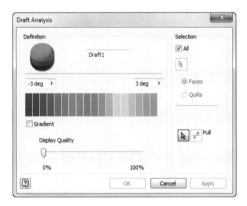

4. Click OK to close the dialog box and display the results of the analysis, as shown in Figure 9.17.

5. After inspecting the results on the part, find the Analysis folder in the Browser, right-click, and uncheck Analysis Visibility in the context menu.

FIGURE 9.17 Use draft analysis to see whether your part can be released from an injection mold.

 TIP This tool does not analyze the shrinkage of material or do any material flow analysis. Those tools are available in more advanced versions of Inventor.

THE ESSENTIALS AND BEYOND

Working with the specialized plastic part features, you learned how a common feature can have its options and variables built into a simple dialog box.

Multibody parts have many applications beyond plastics. They can be used any time more than one part needs to share a common shape.

ADDITIONAL EXERCISES

▶ Apply the groove side of the lip to the back body in c09-04.ipt.

▶ When you are going to apply edge fillets between features, try using rule fillets instead.

▶ Use the Rib tool to create ribbing at an angle or across openings by defining a depth.

▶ The Emboss tool can also cut material. Try doing the exercise for the Emboss tool using a cutting option instead.

CHAPTER 10

Working with Sheet Metal Parts

Sheet metal fabrication follows rules. For bending operations, certain things are not allowed. For example, you can't fold material if you cannot cut the part in the flat pattern, and material comes in certain standard thicknesses.

It's entirely possible to construct sheet metal parts using solid modeling tools, such as Extrude and Revolve, but working with specialized tools eliminates steps, and because these tools work with styles, changing the style that the part is based on updates the part.

In this chapter, you will create different parts using the sheet metal tools of the Autodesk® Inventor® software. The operation of these tools is essentially similar to other solid modeling tools, so they should be easy to learn.

▶ **Defining sheet metal material styles**

▶ **Building sheet metal components**

▶ **Preparing the part for manufacture**

▶ **Documenting sheet metal parts**

Defining Sheet Metal Material Styles

Like the dimension and layer styles you explored in Chapter 5, "Customizing Styles and Templates," sheet metal defaults are styles used to establish consistency between components. You can create these styles within a part (with a scope limited to that part) or save them into a template to be available as you create new parts.

1. Make certain that the 2013 Essentials project file is active, and then open the file c10-01.ipt from the Parts/Chapter 10 folder.

2. Click the Sheet Metal Defaults tool (Figure 10.1) on the Setup panel of the Sheet Metal tab, and click the edit icon next to the Sheet Metal Rule drop-down.

3. When the Styles and Standard Editor dialog box opens, make sure the Default_mm style is active, and click the New icon at the top.

4. In the New Local Style dialog box, enter the name **Aluminum 3 mm**, and click OK.

5. In the dialog box, change the Material drop-down to Aluminum-6061.

6. Set Thickness to 3 mm and the Miter/Rip/Seam Gap value to Thickness/2.

7. Set Flat Pattern Punch Representation to Center Mark Only. See Figure 10.2.

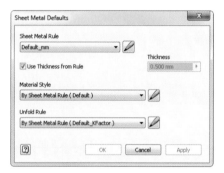

FIGURE 10.1 The Sheet Metal Defaults dialog box controls whether you're basing your part on a predefined rule or what settings to use if you're not.

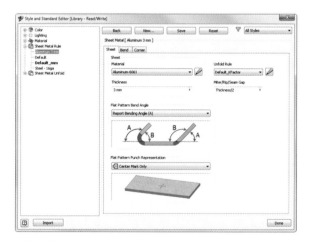

FIGURE 10.2 On the Sheet tab of the Styles and Standard Editor, you define the thickness and material of the sheet.

8. Click the Bend tab, and change Relief Shape to Tear and Bend Radius to 2 mm, as shown in Figure 10.3.

9. Click the Corner tab. Change 2 Bend Intersection Relief Shape to Linear Weld.

10. Change the Relief Shape value for 3 Bend Intersection to Full Round. See Figure 10.4.

11. Click Save, and then double-click the Aluminum 3 mm rule (on the left) to make it active.

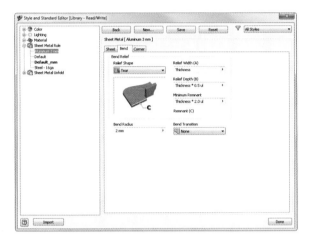

FIGURE 10.3 Use the Bend tab to control the radius of the bend and how edges will intersect with the sheet.

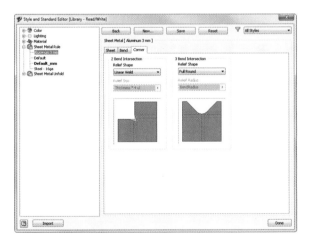

FIGURE 10.4 The Corner tab regulates how material will be removed from the corner or whether it will be left in.

By creating a series of sheet metal defaults and saving them in a template, you can save time by not re-creating the styles over and over.

Building Sheet Metal Components

Even though the Inventor sheet metal tools are limited to components that can be made with break press operations, the variety of components that can be made with this process is enormous. As a result, there are a large number of sheet metal tools to accommodate the different features that are commonly needed.

Creating a Basic Face

The Face tool looks much like the Extrude tool, but with sheet metal features, there is no need to input a thickness from the material because that is determined by the sheet metal default style.

Face

1. Verify that the 2013 Essentials project file is active, and then open c10-02.ipt from the Parts/Chapter 10 folder.

2. Start the Face tool from the Create panel of the Sheet Metal tab.

3. There is more than one closed profile, so no profile is automatically selected. Select the larger portion of the sketch, as shown in Figure 10.5.

4. Click OK to create the face.

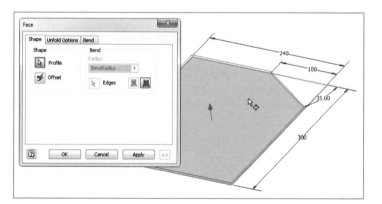

FIGURE 10.5 Select the portion of the sketch from which to create the face.

 TIP A face feature is frequently the base feature for sheet metal parts. You should try to use it to define the overall shape of your part. Face features can be added to a part at any time.

Adding Sides to the Part

Flanges are typically features that are bent from the base. The Flange tool includes nearly every conceivable option for creating your features.

Certification
Objective

Flange

1. Verify that the 2013 Essentials project file is active, and then open c10-03.ipt from the Parts/Chapter 10 folder.

2. Locate and start the Flange tool in the Create panel of the Sheet Metal tab or the marking menu.

3. On the Shape tab, click the Loop Select mode, and click the front face of the part to highlight all edges and get a preview of the feature.

4. Set the Height Extents value to 40. If the flange preview is not in the same direction as Figure 10.6, hold the Shift key, deselect the lower edge, and then reselect the edge of the top face.

TIP When selecting edges for placing flanges, it's important to be aware of which edge you're selecting. The height is based on the edge you select. If you want the height measured from an outside face, you can reverse the direction of the flange to pass back through an existing face.

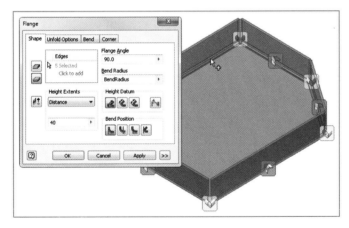

FIGURE 10.6 Select the portion of the sketch from which to create the face.

5. Click Apply to create the new flanges.

6. Switch to Edge Select mode, and set Height Extents to 20.

7. Click the inside edges of the previously placed flanges, as shown in Figure 10.7. If you cannot select an edge, click next to 0 Selected to enable selection mode.

8. Select the Corner Edit tool at the top-left corner of the part.

9. When the Corner Edit dialog box opens, click the Corner Override check box, and select Reverse Overlap from the drop-down. See Figure 10.8.

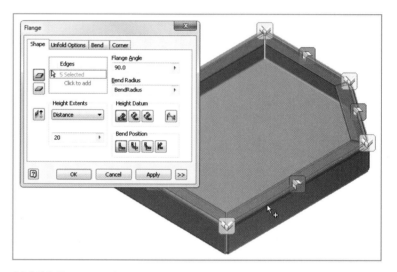

FIGURE 10.7 Corners that would otherwise overlap are automatically mitered.

FIGURE 10.8 The automatic corner treatments can be overridden.

10. Click OK to accept the corner edit and then OK to place the flanges.

11. Start the Flange tool again, and click the front edge of the new flange on the angled face. (See Figure 10.9.)

12. Expand the dialog box to see additional options, set Width Extents Type to Offset, and change the offset values to 10 mm.

13. Change Flange Height to 15 mm, as shown in Figure 10.9, and click OK to place the feature.

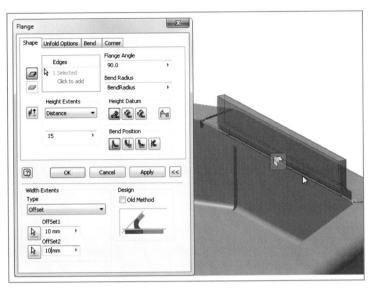

FIGURE 10.9 Adding a flange with the ends offset from the edge length

The angle of a new flange and how it is developed in the footprint of the existing parts are all important options worth experimenting with when using the Flange tool.

Building from the Middle

At times, you might need to build faces in space or from projections from other parts in the assembly. Faces can be created in space with no connection to the rest of the part, but eventually they will need to be attached to the rest of the part. The Bend tool can help you do that using more than one workflow.

Certification
Objective

1. Verify that the 2013 Essentials project file is active, and then open c10-04.ipt from the Parts/Chapter 10 folder.

2. On the Sheet Metal tab, start the Bend tool from the Create panel.

3. Click one of the long edges of the red Face feature and then an edge on the nearest flange of the main body.
   The preview shows a feature that bridges the selected edges and cuts into the flange on the main part.

4. Click the Full Radius option in the Bend dialog box. Notice that the preview shows a feature that is tangent to both faces.

5. Return to the 90 Degree option, and click the Flip Fixed Edge option. Your screen should look like Figure 10.10.
   This bases the geometry on the selected edge of the main part.

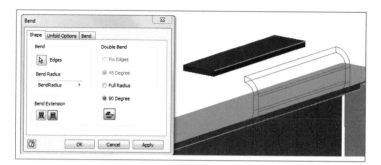

FIGURE 10.10 The Bend tool builds geometry between features.

6. Click OK to create the new feature.

 TIP A face can also be used to add geometry to a part when the shape is too complex for a flange feature.

7. Move the end-of-folded marker to the bottom of the Browser to expose a sketch.

8. Start the Face tool from the Create panel.

9. When the face preview appears, click the Edges selection icon, and click the same edge on the main body as you did in step 3.

10. After the dialog box expands, click the 90 Degree Double bend, and click the Flip Fixed Edge option.

11. Click OK to generate the feature; Figure 10.11 shows the result.

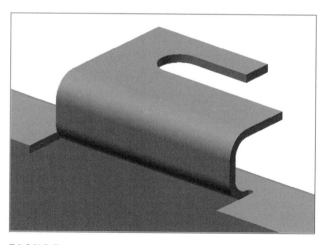

FIGURE 10.11 The generated Bend feature

You can create these bend features from an existing feature or while creating a new one. The other bend options, such as limiting the bend to 45 degrees or using Flip Fixed Edge, can build geometry that could take several features to make without them.

Using an Open Profile

To create a long part with a consistent profile, you could build a series of flanges attached to one another, but using the Contour Flange tool is a much simpler way to go.

1. Verify that the 2013 Essentials project file is active, and then open c10-05.ipt from the Parts/Chapter 10 folder.

2. Start the Contour Flange tool from the Create panel on the Sheet Metal tab.

3. Select the profile, and the preview appears. Expand the dialog box if need be, and set the Distance value to 300.

4. Change the direction for the feature as shown in Figure 10.12.

5. Click OK to create the feature.

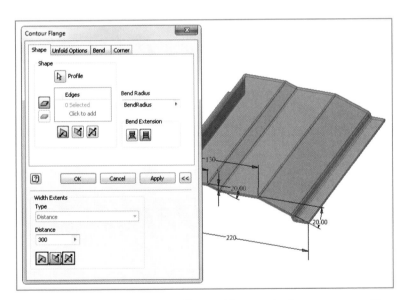

FIGURE 10.12 The contour flange adds geometry from an open profile.

Without this tool, creating the part you've just built would have taken a face and seven flanges. Contour Flange adds the band radii at the corners of the sketch. You can build curves into the sketch, and they will be maintained in the feature.

Adding Library Features Around Bends

A very challenging problem is dealing with features that transition across bends. Another challenge is working with features that you will use repeatedly. Punch tools are a common need in sheet metal, and you can define them and save them to a library for reuse.

1. Verify that the 2013 Essentials project file is active, and then open c10-06.ipt from the Parts/Chapter 10 folder.

2. Start the Unfold tool from the Modify panel of the Sheet Metal tab.

3. When the Unfold dialog box opens, click the top face of the part for the stationary reference.

4. Once the stationary reference is selected, the bends of the part will be highlighted. Click the bend shown in Figure 10.13, and then click OK.

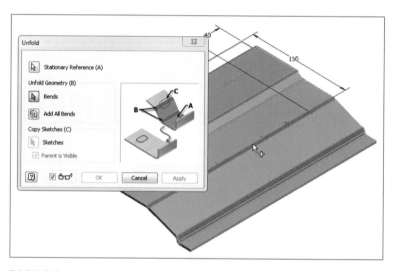

FIGURE 10.13 Removing one of the folds in the part

An Inventor Punch Tool feature can be placed only on a flat surface. By unfolding a portion of the part, you've created a flat surface large enough on which to place the feature.

5. Click the Punch Tool icon in the Modify panel.

6. In the Punch Tool Directory dialog box, find the Square Emboss.ide feature, and double-click it to start placing it on the part.

Because there was a pattern of hole centers in the sketch, four previews appear for the Punch tool.

7. Switch to the Size tab of the PunchTool dialog box, and set the Length value to 60 mm and the Height value to 4 mm, as shown in Figure 10.14. (The preview will not update.)

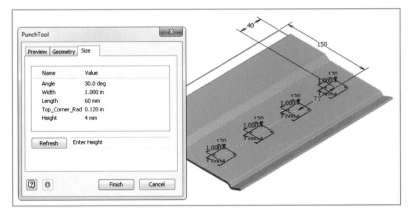

FIGURE 10.14 Placing the punch tools on the part

8. Click Finish to build the features.

Now, you need to restore the part to its original shape. This requires the punch tools to be bent with the other geometry.

9. Click the Refold tool in the Modify panel.

10. Once again, use the top face as the stationary reference, and click the only unfold feature that highlights.

11. Click OK to restore the shape of the part, as shown in Figure 10.15.

12. Move your end-of-folded marker to the bottom of the list to expose the cut sketch.

13. Edit the cut sketch by double-clicking it in the Browser.

FIGURE 10.15 The part's bend restored

Project
Flat Pattern

14. On the Format panel, click the Construction override.

15. Expand the Project Geometry tool in the Draw panel of the Sketch tab, and click Project Flat Pattern.

16. Click the face that the rectangle of the sketch overhangs.

17. Add a 15 mm dimension between the end of the rectangle and the end of the projection of the space between the punch tools, as shown in Figure 10.16.

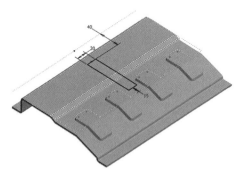

FIGURE 10.16 Projecting the flat pattern makes it possible to dimension a position after it will be bent.

18. Finish the sketch.

Cut

19. From the marking menu or the Modify panel of the Sheet Metal tab, start the Cut tool.

20. In the Cut dialog box, select the Cut Across Bend check box, and click OK to see the finished part in Figure 10.17.

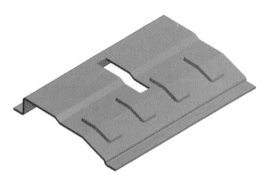

FIGURE 10.17 The position of the cut can be controlled on multiple faces.

Measuring the distance between the new cut and the edge of the face on which it ends shows that the distance includes the transition across the two bends.

Exploring an Advanced Open Profile Tool

The Contour Flange tool uses an open profile to generate a complex shape normal to the sketch. You can use the Contour Roll tool to revolve a profile around an axis to develop a sheet metal part.

1. Verify that the 2013 Essentials project file is active, and then open c10-07.ipt from the Parts/Chapter 10 folder.

2. Start the Contour Roll tool from the Create panel of the Sheet Metal tab. The Contour Roll tool detects the single sketch and selects it, but you still need to select an axis.

3. Select the centerline to which the dimension of 200 is attached.

4. When the preview appears, change Rolled Angle to 30 deg, as shown in Figure 10.18, and click OK to generate the feature.

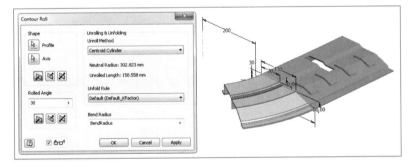

FIGURE 10.18 Bending the open profile around an axis

The Contour Roll tool is interesting, and it can create components that would be difficult to calculate using other tools.

Building Transitions in Sheet Metal

There are many uses for sheet metal components that change from one shape to another. Defining these shapes for manufacturing has been a longtime challenge for designers.

1. Verify that the 2013 Essentials project file is active, and then open c10-08.ipt from the Parts/Chapter 10 folder.

The profiles of the lofted flange do not have to be parallel to one another to create the feature.

2. On the Sheet Metal tab, find the Lofted Flange tool in the Create panel, and start it.

3. Click the rectangle and the circle sketches in the Design window.

4. After the preview appears, click the Die Formed option in the dialog box to see how the part would look using that process.

5. Return to the Press Brake option, and set the Chord Tolerance value to 3 mm, as shown in Figure 10.19.

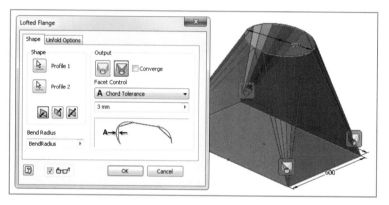

FIGURE 10.19 Being able to see how the transition will be fabricated can help guide decisions.

The model updates to show the number of facets that need to be bent to create the part with this level of precision. Additional options for Facet Control include the angle between facets and the distance between points on the long edge.

6. Click OK to create the feature.

It will not be possible to create a flat pattern of the component as it is now. You will need to create a gap using the Rip feature to be able to unfold the part.

7. Start the Rip tool from the Modify panel.

8. In the dialog box, use the Rip Type drop-down to set the value to Point To Point.

9. Set the model to the Home view, and select the large triangular face aligned with the bottom of the ViewCube as the rip face.

10. Make the start point the top tip of the face and the end point the middle of the bottom edge, as shown in Figure 10.20.

11. Click OK to create the rip in the part.

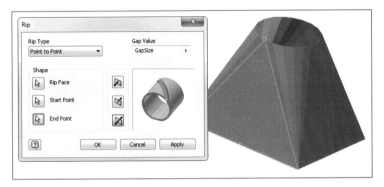

FIGURE 10.20 Making it possible to calculate the flat pattern

The exercise noted the eventual need to create a flat pattern. Later in this chapter, I will cover the tools for creating a flat pattern.

Working with Existing Designs

If you or your employer has a history of making sheet metal components, it is likely that you already have some flat patterns that work well for your designs. It is possible to use those existing .dwg or .dxf files to create the 3D part and even to add features to it.

1. Verify that the 2013 Essentials project file is active, and then open c10-09.ipt from the Parts/Chapter 10 folder.

2. Edit Sketch 1 in the Browser.

3. Click the Insert AutoCAD tool on the Insert panel of the Sketch tab.

4. Open c10-09.dwg from the Parts/Chapter 10 folder.

5. When the Import Destination Options dialog box opens, click the Next button to keep all layers selected, and then make sure the Constrain End Points option is selected and click Finish.

6. When the geometry appears in the sketch, finish the sketch.

7. Select Zoom All so you can see the entire sketch.

8. In the Create panel, click the Face tool, and click all four portions of the sketch.

9. Click the Offset tool to change the direction in which the face feature will be developed, and click OK to create the face.

10. In the Browser, expand the face feature, right-click the sketch, and click Share Sketch in the context menu.

Now the sketch is visible, and you can use the bend lines from the flat pattern to fold the material into a 3D form.

11. Start the Fold tool on the Create panel.

12. For the first Bend Line selection, click the red line beside the number 1.
     This displays a glyph of a straight arrow showing which side of the bend will be folded and a curved arrow to show the direction of the fold.

13. Click Apply to fold a small portion of the part.

14. For the second edge, click the line next to the number 2, and select Apply.

15. Click the edge near the number 3. Notice that the glyph indicates that it will fold up the wrong portion of the part.

16. Click the Flip Side icon to make sure that the correct geometry (Figure 10.21) will be folded, and click OK.
     The component is now suitable for building new drawings or developing new features. Changes to the sketch will modify the folded component as expected.

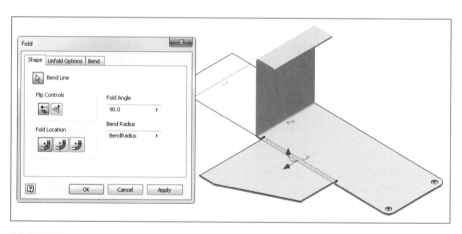

FIGURE 10.21 Folding a flat portion of a face around a line

Adding the Finishing Touches

Creating the big portions of a sheet metal part has been covered so far, but the details are also important. Adding chamfers, fillets, and hems to the metal improve safety for the customer and add strength to the part.

1. Verify that the 2013 Essentials project file is active, and then open c10-10.ipt from the Parts/Chapter 10 folder.

2. Zoom in on the top-left corner of the folded part.

3. Start the Corner Chamfer tool from the Modify panel.

Once the dialog box is open, you can select edges, but unlike the normal Chamfer, this tool only selects the edges created normal to the sheet so that the feature can be cut from material using the flat pattern.

4. Click the corner shown in Figure 10.22, set the value to 5 mm, and click OK to place the chamfer.

FIGURE 10.22 Corner chamfers can be cut only in the flat pattern.

5. Click the Corner Round tool on the Modify panel.

6. Select the edge shown in Figure 10.23, set the radius to 5 mm, and click OK to place the fillet.

Another useful option in the Corner Round tool is a feature selection option, which can find all the available edges with one click.

7. Find the Hem tool in the Create panel, and start it.

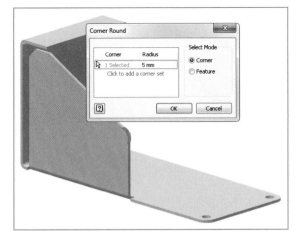

FIGURE 10.23 A corner round follows the same limitations as the corner chamfer.

8. After the dialog box opens, click the long edge of the short face (Figure 10.24).

9. Use the Type drop-down to set the option to Teardrop.

10. Set the Radius value to 0.8 mm, as shown in Figure 10.24, and click OK to place the hem.

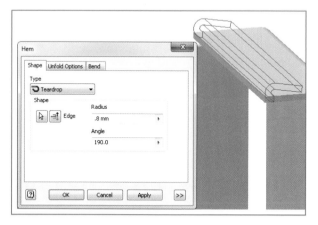

FIGURE 10.24 Hems can be complex features but are added easily.

The Hem tool's four options should cover almost any requirement for fabrication. Now, you can focus on making your components suitable for the manufacturing process.

Preparing the Part for Manufacture

In some cases, you might choose to send data for the finished part to an external resource and allow that person to calculate the material needed to create the part. Inventor has the capability to develop a flat pattern for a part. You can also edit flat patterns to add material that isn't included in the folded part.

Building the Flat Pattern

Certification Objective

A flat pattern for a part includes calculations for the stretch of the material as it is formed. The type, thickness, and bend radius can affect this calculation. It is also common for companies to build in extra material to cause deformation, and this is done by modifying the flat pattern.

1. Verify that the 2013 Essentials project file is active, and then open c10-11.ipt from the Parts/Chapter 10 folder.

2. Start the Create Flat Pattern tool from the Sheet Metal tab's Flat Pattern panel. Figure 10.25 shows the resulting flat pattern.

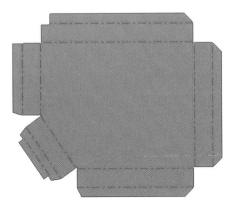

FIGURE 10.25 The flat pattern of the part

Take a moment to look at the Browser. The flat pattern appears in the Design window, but the Browser is also updated to show the flat pattern state.

3. Start the Distance tool on the Measure panel of the Tools tab.

4. Click the far-left and far-right edges of the flat pattern, and make note of the measured distance.

5. Return to the Flat Pattern tab, and click Go To Folded Part in the marking menu or the Folded Part panel.

6. Start the Sheet Metal Defaults tool from the Setup panel or marking menu, and using the Sheet Metal Rule drop-down, switch to the Steel – 16 ga style.

7. Click OK, and watch the part closely to see it change.
   You should see that the material is thinner and the gaps between edges are smaller, and you will also see that the last flange added has a relief at either end.

8. Return to the flat pattern by double-clicking its icon in the Browser.

9. Make the same measurement as in step 4, and see the difference in the size of the flat pattern.

10. Right-click the flat pattern in the Browser, and click Extents in the context menu. See Figure 10.26 for the results.

11. Close the dialog box, press the F6 key for a home view of the flat pattern, and then zoom in on the corner of the part that is near the bottom of the Design window.

Once a flat pattern is generated, you can right-click it in the browser and use the Save Copy As feature to export it as a DXF, DWG, or SAT solid model.

FIGURE 10.26 The Flat Pattern Extents dialog box reports the material needs of the part.

12. Start the Fillet tool by pressing F (click OK to modify the flat pattern), and set it to a face fillet.

13. Click the two sides of the part shown in Figure 10.27, and set the radius to 5 mm.

14. Click OK to place the fillet in the flat pattern.

FIGURE 10.27 You can add features like fillets to the flat pattern.

The fillet that was added appears on drawings of the flat pattern but does not propagate back to the 3D model. Features added to the flat pattern are specifically for the use and benefit of the manufacturing process.

Documenting Sheet Metal Parts

The ability to create documentation in the drawing view that you've seen with other solid models works just as well for sheet metal components, but they need even more. For sheet metal, it's important to document not only the folded part but also the flat part, and being able to communicate the process for forming it is a great benefit.

Establishing the Process

In the design phase, the flanges, bends, and faces of the part can be built in any particular order. That's not true for producing the part. Inventor has the ability to specify the order in which you want folds to be made in the solid model. The order you specify can then be accessed in the drawing.

1. Verify that the 2013 Essentials project file is active, and then open `c10-12.ipt` from the `Parts/Chapter 10` folder.

2. Activate the flat pattern in the model.

3. Start the Bend Order Annotation tool from the marking menu or the Manage panel of the Flat Pattern tab.

   Bend Order Annotation

   The flat pattern will display numbers that represent the order in which the part would be folded.

4. Click the balloon with the number 6.

5. When the Bend Order Edit dialog box appears, select the Bend Number check box, and change the value to 1. Leave the Unique Number option selected, and click OK.

   This will change the number and the icon to identify it as an override.

6. Change the balloon for 9 to 3, and click OK, as shown in Figure 10.28.

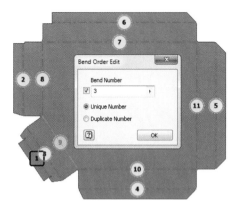

FIGURE 10.28 Like BOM data in the assembly, bend order information is kept in the part file.

Adding this information to the part makes it accessible by anyone who would need to make drawings from your part.

Documenting the Process

Certification
Objective

Sheet metal parts often require two types of drawing. A drawing of the folded part will help quality control and manufacturing verify that the part was built properly. Drawings of the flat patterns can be used to cut the parts out of the raw material.

1. Verify that the 2013 Essentials project file is active, and then create a new drawing using ISO.idw from the Metric Folder of the New File dialog box.

2. Start the Base View tool from the marking menu or the Create panel of the Place Views tab.

3. Click the Open An Existing File icon, and open the c10-13.ipt file from the Parts/Chapter 10 folder.

4. On the Component tab of the Drawing View dialog box, click the flat pattern for the Sheet Metal view, and set the scale to 1:2.

5. Place the view in the middle of the page, right-click, and select OK from the context menu.

6. Switch the Ribbon to the Annotate tab, and click Bend on the Feature Notes panel.

7. Click the bend line in the drawing closest to the top; when the note is placed, press the Esc key, or right-click and click OK in the context menu.

   Clicking and dragging the insertion point of the bend note will place the note on a leader.

8. Start the Table tool from the Table panel on the Annotate tab.

9. When the Table dialog box opens, click the drawing view.

   The dialog box changes when the view is selected and lists the columns that will be shown in a bend table.

10. Click OK in the dialog box, and then place the table on the drawing. See Figure 10.29.

The numbers of the bends are added to the drawing views and correspond to any edits made to the bend order.

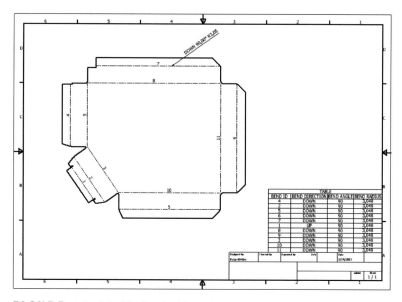

FIGURE 10.29 The bend table and callouts on the drawing of the flat pattern

11. Right-click the drawing view, and select Open to view c10-13.ipt in its own tab.

12. Right-click the flat pattern in the Browser, and click Save Copy As in the context menu.

13. In the Save Copy As dialog box, use the drop-down menu to change the file type to DXF, and then click Save.

14. In the Flat Pattern DXF Export Options dialog box, set File Version to AutoCAD R12/LT 2 DXF.

15. Activate the Geometry tab, select the Merge Profiles Into Polyline check box, and click OK.

This generates a new .dxf file, which is the most common file format for use with two-axis cutting equipment. You can also generate .sat and .dwg files for solid export or direct use with AutoCAD.

THE ESSENTIALS AND BEYOND

Sheet metal tools keep the focus on creating the correct geometry as quickly and easily as possible. Inventor's use of styles to establish the material and its tools that use industry terms make using these tools more natural for anyone who commonly needs to create sheet metal parts. Tools like Contour Flange can simplify the process more by creating many features in one step. Manipulating the flat pattern is a necessity for manufacturing to make sure they can keep using processes that they've used in the past.

ADDITIONAL EXERCISES

▶ Set up templates using the material and sheet gauges you typically use.

▶ Use the Flat Pattern tools on the earlier exercises in the chapter.

▶ Experiment with other Bend options.

▶ Resize and reposition the profiles for the lofted flange and see the update.

CHAPTER 11

Building with the Frame Generator

The specialized tools in Autodesk® Inventor® software help you shortcut arduous and repetitive processes. The design accelerators and plastic part features you worked with in earlier chapters are great examples of this. Another is the Frame Generator. The frames you create with this tool are constructed from dozens of metal shapes contained in the Content Center. Once you place them, you can control how the frame members are joined together, and if the skeleton they're based on changes, the members update as well.

▶ **Creating metal frames**

▶ **Editing metal frames**

Creating Metal Frames

To build a frame, you must have a sketch (2D or 3D) or a solid model to use as a skeleton. You place the frame members by selecting edges or by clicking a beginning point and an end point for the frame element. You can tell the frame member how to position itself along its direction based on nine control points around its section that show how the profile will be oriented to the sketch element. You can also offset the profile a distance from the sketch axis or rotate the section.

Beginning the Frame

The techniques for building the skeleton of the frame are no different from any sketching or part modeling technique covered in previous chapters of this book. For that reason, I will shortcut the process by placing a finished skeleton part into an assembly.

1. Make sure that 2013 Essentials is the active project file, and create a new assembly file using the Standard (in).iam template from the English folder under Templates in the New File dialog box.

Place

2. Use the Place tool on the Component panel of the Assemble tab to add the c11-01.ipt file from the Parts\Chapter 11 folder, as shown in Figure 11.1.

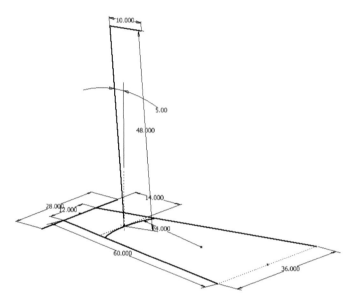

FIGURE 11.1 The frames skeleton consists of a series of parametric sketches.

3. Press the Esc key to quit the Place Component tool.

4. Close the file without saving. For the next exercises, you will use an assembly with the frame already in place.

Whether you load a part file or create a new part in the assembly to use as the skeleton, this part will always be separate from the rest of the frame.

Certification
Objective

Inserting Members on Edges

The process of placing the members requires that you click the geometry on which you want to place them, select a standard of material, and then click the shape to use and the size. You can also select a color or material property for the member.

You will now use a skeleton to begin the process of building your frame.

1. Make certain that the 2013 Essentials project file is active, and then open the c11-01.iam file from the Assemblies\Chapter 11 folder.

Insert
Frame

2. Switch the Ribbon to the Design tab, and start the Insert Frame tool from the Frame panel.

When placing a frame member on an edge, the section might be skewed around the axis. You can enter an angle for the section or use the Align option to select an edge to align to.

3. In the Insert frame dialog box, set Standard to ANSI, Family to ANSI AISC (Square) – Tube, and Size to 3 × 3 × 3/16, with the orientation in the center of the profile, as shown in Figure 11.2.

4. Make sure the Insert Members On Edges Placement option is active, and click all the black lines (and only the black lines) on the skeleton. Your results will look like Figure 11.3.

5. Click OK to generate the frame members; then click OK again to approve the creation of the new files and a third time to approve the names of the new files.

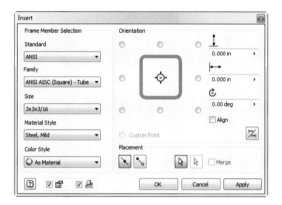

FIGURE 11.2 Setting the type of metal section

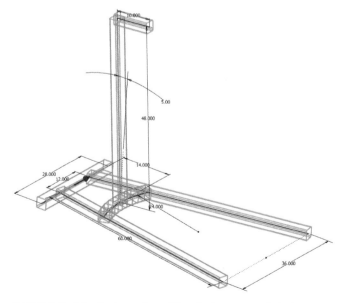

FIGURE 11.3 A preview of the steel components to be placed in the frame

It is possible to change the names and paths of these new files to suit your needs.

Inserting Members Between Points

Rather than creating a lot of extra lines or model edges, you can place the members by selecting a start point and an end point.

It is a good idea to give unique names to the frame subassembly and skeleton when prompted. Filename prompting can also be disabled.

1. Make certain that the 2013 Essentials project file is active, and then open the c11-02.iam file from the Assemblies\Chapter 11 folder.

2. Switch the Ribbon to the Design tab, and start the Insert Frame tool from the Frame panel.

3. In the Insert frame dialog box, set Standard to ANSI, Family to ANSI L (Equal angles) – Angle steel, and Size to L 1.5 × 1.5 × 3/16, keeping the orientation in the center of the profile.

4. Make sure the Insert Members Between Points Placement option is active, and click the end point of the dashed, pink line and the line concealed in the upright member of the frame, as shown in Figure 11.4.

Selection points must be in a sketch. If there is a work point in the model you want to use, you will need to project it into a sketch first.

5. Change the orientation of the profile to the top-center option, as shown in Figure 11.4.

6. Click the Align option and pick the centerline of the 28-inch member. Then set the offset values to -.25 and .55, as shown in Figure 11.4.

7. Click OK to generate the frame members, and then click OK to approve the names of the new file.

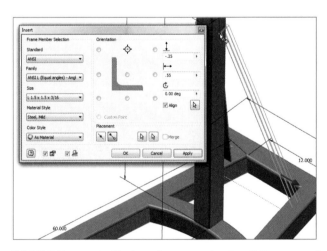

FIGURE 11.4 Placing a frame member by clicking two points

The ability to realign the profile gives you flexibility with open metal shapes.

Inserting Members on Curves

Frame members can be placed on curved edges as well. Note that members placed on curved edges cannot be edited with the frame-editing tools covered in the next section of the chapter. To edit curved frame members, use solid modeling tools like Extrude or Split to remove portions of these components.

1. Make certain that the 2013 Essentials project file is active, and then open the c11-03.iam file from the Assemblies\Chapter 11 folder.

2. Switch the Ribbon to the Design tab, and start the Insert Frame tool from the Frame panel.

3. In the Insert dialog box, set Standard to ANSI, Family to ANSI AISC (Rectangular) – Flat Bar Steel, and Size to 2 ¼ × 16. Keep the orientation in the center of the profile.

4. Make sure the Insert Members On Edges Placement option is active, and click all the yellow lines and arcs on the skeleton.

5. Change the orientation point to be the middle right, set the rotation angle of the section to 90 degrees, and select the Merge check box. Your results will look like Figure 11.5.

   You might need to use a different orientation to achieve the correct results. Use the preview as a guide in all situations.

6. After the preview appears and is positioned correctly, click OK to create the new frame member, and click it again to approve the name for the part.

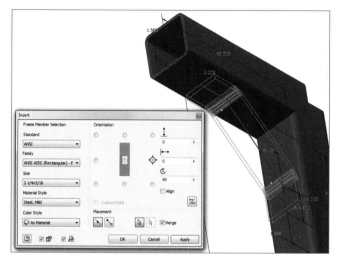

FIGURE 11.5 Building the metal shape across the straight and curved lines

Placing the frame members always follows the same handful of steps. The editing process is also direct. The tools are dialog box–driven and typically require only a couple of selections.

Editing Metal Frames

There are only a few editing tools for the Frame Generator, but they are more than enough to make the type of changes typically needed in the process of designing metal frames. Some of the tools allow for building in gaps for welding operations.

Any change in the skeleton will update the frame members as well as the edits made to them.

Defining Joints with the Miter Tool

A common joint for metal frames is an equal cut on both parts. A gap can be included in the miter equally divided between both parts, or it can be taken from one side or the other.

1. Make certain that the 2013 Essentials project file is active, and then open the c11-04.iam file from the Assemblies\Chapter 11 folder.

2. Switch the Ribbon to the Design tab, and start the Miter tool from the Frame panel.

    When the Miter dialog box opens, you will be prompted to click two frame members.

3. Click the blue beam for the first option and the red one for the second.

4. Set the gap to 0.130 in, as shown in Figure 11.6, and click OK to create the miter.

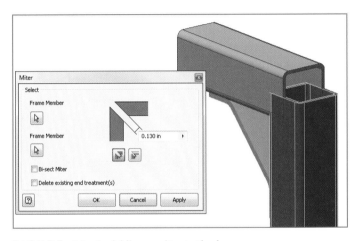

FIGURE 11.6 Adding a miter to the frame

Adding the miter changes the length of both frame members. Any end treatment applied to a frame member can be removed.

Changing an Edit and the Trim to Frame Tool

The Trim to Frame end treatment will take two members and set one to butt against the other and then limit the length of the first member to the width of the second.

The existing frame has a miter applied to two sections. You need to change it to a Trim to Frame joint.

1. Make certain that the 2013 Essentials project file is active, and then open the c11-05.iam file from the Assemblies\Chapter 11 folder.

2. Switch the Ribbon to the Design tab, and start the Trim to Frame tool from the Frame panel.

   When the Trim to Frame dialog box opens, you will be prompted to click two frame members. The first, represented in the dialog box in blue, overlaps the second, represented in yellow.

3. Click the orange section for the first option and the nearly vertical section mitered to it as the second selection, as shown in Figure 11.7.

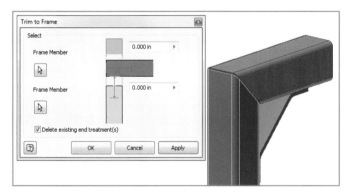

FIGURE 11.7 Modifying the miter to be an overlap

4. Select the Delete Existing End Treatment(s) check box, and click OK to update the model, as shown in Figure 11.8.

This removes the miter end treatment and then sets the orange member to overlap the green.

FIGURE 11.8 Trim to Frame will accommodate the sizes of both frame members for a clean overlap.

The Trim/Extend Tool

When frame members are placed on the skeleton, they're often overlapping or might fall short of other members they need to butt against. This end treatment allows you to select several members and terminate them at a selected face, whether it means trimming or extending them.

1. Make certain that the 2013 Essentials project file is active, and then open the c11-06.iam file from the Assemblies\Chapter 11 folder.

2. Switch the Ribbon to the Design tab, and start the Trim/Extend tool from the Frame panel.

3. When the dialog box opens, click the two longest members at the base. Once they're selected, right-click and click Continue from the context menu to select the trimming face.

4. Click the face that the two selected segments intersect, as shown in Figure 11.9, and then click OK.

You must select model faces for the trimming surface. The model face can be a face on the skeleton if it is a solid model.

Creating Notches

For more complex joints between frame members, you can cut one member with the shape of another.

1. Make certain that the 2013 Essentials project file is active, and then open the c11-07.iam file from the Assemblies\Chapter 11 folder.

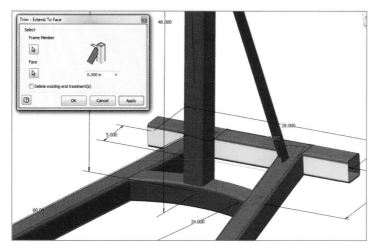

FIGURE 11.9 Select a face to which to match the frame members.

2. Switch the Ribbon to the Design tab, and start the Notch tool from the Frame panel.

From the Notch dialog box, you select a frame member to cut first and then the member with which to cut it. These will be highlighted in blue and yellow as they are selected.

3. Click the yellow part first and then the red part for the cutting profile, as shown in Figure 11.10.

4. Click OK to create the feature. See the results in Figure 11.11.

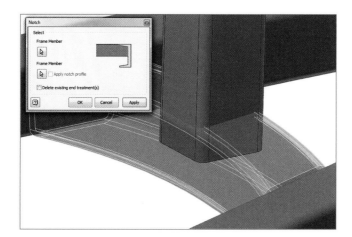

FIGURE 11.10 The notch will update with any changes to the geometry.

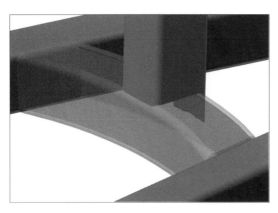

FIGURE 11.11 A notch will remove an exact shape from the selected member.

TIP In cases where a pure notch creates too detailed a cut, it's advisable to do a traditional extruding cut or create a custom coping profile.

The Lengthen/Shorten Tool

The length of a placed frame member is based on the length of the selected edge or the distance between the selected points. The Trim/Extend tool is useful if you have a face to trim or to which to extend. This tool is for changing the length of a member in open space or when you don't want to use a face on another member without changing the skeleton itself.

1. Make certain that the 2013 Essentials project file is active, and then open the c11-08.iam file from the Assemblies\Chapter 11 folder.

 Lengthen/Shorten

2. Switch the Ribbon to the Design tab, and start the Lengthen/Shorten tool from the Frame panel.

   The tool can add length to one end or equally on each end of the selected member.

3. In the Lengthen – Shorten Frame Member dialog box, set the extension length to 4.5 in, and leave the Extension Type set to Lengthen – Shorten Frame Member At One End.

4. For a single direction, the selection is click-sensitive. You need to click the end that will change. Click the end of the yellow piece that is in the open (Figure 11.12), and click OK.

If you need to add length to an end that has an existing end treatment, you will need to delete the end treatment first or select the Delete Existing End Treatment(s) option.

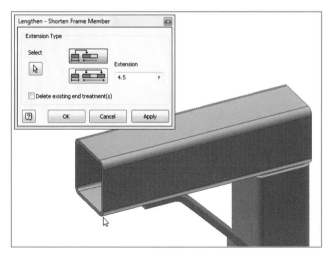

FIGURE 11.12 Selecting the open end of the frame will add length to it.

The Change Tool

Sometimes, you discover that a frame as originally designed needs to be changed. The frame members might need to be lighter or heavier or repositioned. The orientation, size, and family of any frame member can be changed. Using the Change tool, you will select a single or multiple members that are presently the same size. You can then edit the selected members with the same options used to place them.

1. Make certain that the 2013 Essentials project file is active, and then open the c11-09.iam file from the Assemblies\Chapter 11 folder.

2. Switch the Ribbon to the Design tab, and start the Change tool from the Frame panel.

3. When the Change dialog box opens, click the red frame member in the base of the frame. After a short time, its properties will appear in the dialog box.

4. Change the family of the part to ANSI AISC (Rectangular) – Tube and the size to 5 × 3 × 3/16 (Figure 11.13); then click OK to update the component.

5. Click Yes in the dialog box to approve the change to the family of the component, and click OK to create the new part.

6. Orbit the part to see the effect the change has in the assembly, as shown in Figure 11.14.
   The end treatments of the member will be updated even if it's curved.

You can also change the standard for a frame element.

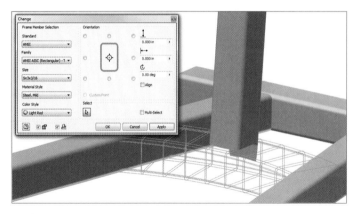

FIGURE 11.13 Changing the family of a frame member

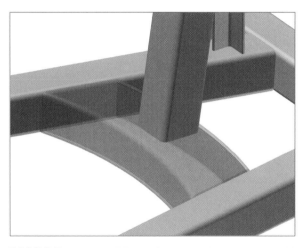

FIGURE 11.14 Edits to frame members are updated by changes to the frame sizes

For all but radical changes to the family of a frame member, the end treatments should be maintained. This is also true for most cases when the skeleton of the frame is changed.

Changing the Frame Skeleton

The great advantage of using the Frame Generator over creating individual parts based on extruded profiles is the association between the frame members and the skeleton of the frame. A change to the skeleton updates the members of the frame rather than requiring you to edit individual parts one at a time.

1. Make certain that the 2013 Essentials project file is active, and then open the c11-10.iam file from the Assemblies\Chapter 11 folder.

2. Expand the c11-03:1 part in the Browser.

3. Right-click Sketch2, and turn on visibility in the context menu.

4. Locate the selection filter drop-down in the Quick Access toolbar, and set the priority to Select Sketch Features.

5. Double-click the 48.000 dimension to edit it. Change the value in the mini dialog box to 54, and click the check mark to update the dimension.

6. Then change Angle Dimension from 5 deg to 10 deg (Figure 11.15), and click the check mark to complete the edit.

7. In the Quick Access toolbar, click the update icon to rebuild the assembly.

8. Turn off the visibility of Sketch 2, and inspect the completed frame assembly, shown in Figure 11.16.

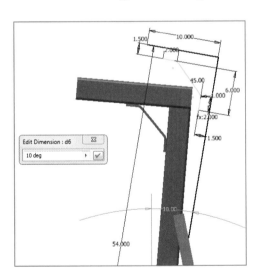

FIGURE 11.15 Editing the parametric
dimensions of the skeleton

The simplicity of the Frame Generator tools and the associativity of the generated frame with the skeleton make metal frames easy to create and maintain. If you have experience creating these types of frames, the tools should help you be more productive immediately.

FIGURE 11.16 The completed frame created with the Frame Generator

THE ESSENTIALS AND BEYOND

The Frame Generator tools open a new way of constructing the frames that are so common in machine design. Placing the frame members directly from the Content Center saves you the work of having to sketch the shapes and sizes of the parts. Using tools such as Trim/Extend, Trim to Frame, and Miter can save you hours of editing on a simple change. Using this workflow also bypasses the placement of the assembly constraints while maintaining the advantages of separate parts for detailing needs.

ADDITIONAL EXERCISES

▶ Make other changes to the skeleton model to see the effects.

▶ Try changing components to a non-ANSI standard shape.

▶ Change the material of the frame members to see how it affects the mass properties of the assembly.

▶ Experiment with changing closed profile members to ones with open profiles and see how the end treatments behave.

CHAPTER 12

Working in a Weldment Environment

A weldment is a specialized assembly. Like the sheet metal tools, the tools that the Autodesk® Inventor® software provides for working with weldments are built around the workflows and demands of a production environment. A weldment is an assembly that is regarded as a single, inseparable part.

Building the weldment as an assembly first makes it easy to understand how the components need to be positioned and what components are included in the weldment. The weldment environment also understands the ramifications weldments have on BOM structure and how to handle the weldment as a single component and document its members when needed.

Welds that are added to the weldment exist only in the environment and do not require you to maintain more files.

▶ **Converting an assembly**

▶ **Calculating a fillet weld**

▶ **Preparing to apply weld features**

▶ **Applying weld features**

▶ **Adding machined features to the weldment**

▶ **Documenting welds and weldments**

Converting an Assembly

There are a few special environments or shells within an assembly, and in order to access the weldment tools for Inventor, you must convert your assembly to a weldment.

Certification Objective

1. Make certain that the 2013 Essentials project file is active, and then open c12-01.iam from the Assemblies\Chapter 12 folder.

Convert to
Weldment

2. Click the Convert to Weldment tool on the Convert panel of the Assemble tab.

3. Click Yes when prompted that you will not be able to revert this data to a basic assembly.

   After you confirm the conversion, the Convert To Weldment dialog will appear for you to input specific information about the standards and structure of the weldment.

4. Set Weld Bead Material to Welded Steel Mild using the drop-down.

5. Click OK to add the Weld tab to the Ribbon (Figure 12.1), and populate the Browser with a Preparations, Welds, and Machining group.

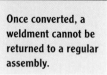

Once converted, a weldment cannot be returned to a regular assembly.

FIGURE 12.1 The three additional states of a weldment

The assembly is still an assembly; it will just be treated differently in the BOM and have access to new tools.

Calculating a Fillet Weld

The design accelerators covered in Chapter 8, "Advanced Assembly and Engineering Tools," focused on tools that generated components. There is another class of tools associated with design accelerators; they are the calculators. The Inventor Weld Calculator tools can calculate several types of welds and solder joints.

Weld
Calculator

1. Make certain that the 2013 Essentials project file is active, and then open c12-02.iam from the Assemblies\Chapter 12 folder.

Fillet Weld Calculator (Plane)

2. Expand the Weld Calculator tool in the Weld panel on the Weld tab.

3. Click the Fillet Weld Calculator (Plane) tool.

   The goal is to keep the weld at the base of the blue arm less than 7 mm. The calculator helps you determine whether this is reasonable.

4. Set the weld form to an all-around rectangle.

5. Set the weld loads to be the bending force parallel with the neutral axis of the weld group.

6. Set the dialog options as follows:

- ▶ Bending Force Value = 12000 N

- ▶ Force Arm length = 365 mm

- ▶ Weld Height = 5 mm

- ▶ Weld Group Height = 70

- ▶ Weld Group Width = 70

- ▶ Joint Material = Structural Steel SPT360

7. Click Calculate to get the results.

The calculator tells you that with the default safety factor of 2.5, you would need a weld bead of just over 5 mm, but the stresses are too high.

8. Change the Weld height to 6 mm, and click Calculate again to find a weld size that will work. Figure 12.2 shows the results.

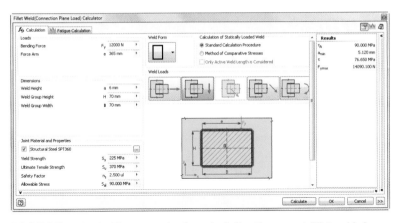

FIGURE 12.2 The properties for calculating the correct fillet weld size

Now that you've calculated the weld, you can apply it with confidence that it is sound. This and the other weld calculators can be used at any time to create a new weld or guide your edits to an existing weld.

Preparing to Apply Weld Features

Preparations are features that are applied to a part after it has been cut but before the weld is applied. These features do not appear in the source part file, because a part might be placed in different assemblies and need different treatments for each. A preparation can be an extrusion, revolved feature, hole, fillet chamfer, or sweep. The most common is the chamfer.

1. Make certain that the 2013 Essentials project file is active, and continue using or open c12-02.iam from the Assemblies\Chapter 12 folder.

Preparation

2. To access the Preparation tools, click the Preparation tool in the marking menu or the Process panel of the Weld tab.
    This step makes tools available in the Ribbon and also changes the Browser to show that these changes are added under the Preparations folder, not to the components in the assembly.

3. Click the Chamfer tool on the Weld tab in the Preparation And Machining panel.

4. Set the distance to 3 mm, and click the three edges highlighted in Figure 12.3.

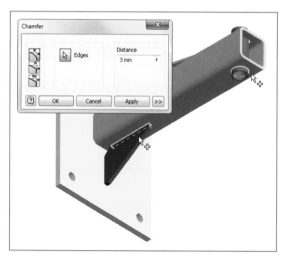

FIGURE 12.3 Adding a chamfer to edges to prepare for the weld

5. Click OK to add the chamfers to the components.

6. Click Return from the Ribbon to leave the Preparation tools. Figure 12.4 shows the results.

Now that you've added the preparation features you need, you can begin to add some welds to the assembly.

FIGURE 12.4 The chamfers appear in the weldment but not the individual parts.

Applying Weld Features

Three types of welds are available: fillet, groove, and cosmetic. In the next set of exercises, you will apply all of them and use a couple of options as well.

Adding a Fillet Weld and the End Fill Tool

Fillet welds are applied by selecting a minimum of two faces. There are two selection sets, but you can add as many faces to each set as you need to define the feature.

1. Make certain that the 2013 Essentials project file is active, and then open c12-03.iam from the Assemblies\Chapter 12 folder.

2. Access the weld tools by clicking the Welds icon from the Process panel on the Weld tab or the marking menu.

Welds

3. When the tools appear in the Weld panel of the Weld tab marking menu, click the Fillet tool.

Fillet

4. In the Fillet Weld dialog, select the Chain check box, and then click the exterior surface of the blue tube.

5. Click the icon for selection set 2, and then select the face on the yellow plate against which the tube butts.
   A preview of the fillet weld appears at the default value of 10 mm or the last size you used.

6. Set the first fillet value to 6 mm per the weld calculation done previously in this chapter.

7. Click Apply to generate the fillet weld.

8. Deselect the Chain option.

9. For the first selection, click the face on the blue tube adjacent to the green gusset.

10. Switch to the second selection set, and click the triangular side and the chamfer surface shown in Figure 12.5.

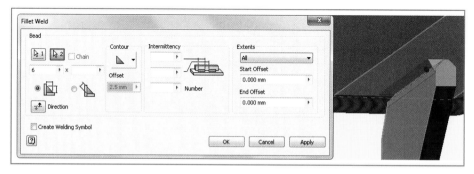

FIGURE 12.5 A weld filling the gap left by a chamfer

11. Click OK to place the weld.
    Because you selected the side and chamfer face on the gusset, the weld fills in all of the space. Selecting only the blue tube and the side of the gusset could still generate a weld, but it would leave a space. One last touch is needed to make the weld look better.

12. In the Weld panel of the Weld tab, find the End Fill tool, and start it.
    The bare ends of the weld bead should highlight. If they do not, click them.

13. Click one of the ends of the bead, and an image will be added to it, as shown in Figure 12.6.

14. Press the Esc key to end the End Fill tool.

Now you've added fillet welds to the assembly. These welds do not require any new part files to maintain them in the assembly.

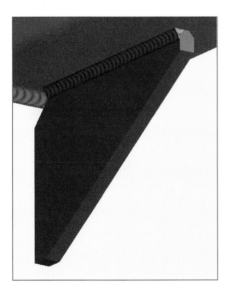

FIGURE 12.6 Adding an end fill improves the appearance of the welds.

Adding a Groove Weld

Groove welds are used for filling gaps between parts. The direction of the gap to be closed is established in the dialog box. There is another interesting option on which this exercise will focus—the radial gap.

1. Make certain that the 2013 Essentials project file is active, and then open c12-04.iam from the Assemblies\Chapter 12 folder.

2. Access the weld tools by clicking Welds from the marking menu or from the Process panel on the Weld tab.

3. Click the Groove Weld tool from the Weld panel.

4. Click the exterior cylindrical face of the red part for Face Set 1.

5. Click the selection icon of Face Set 2, and then click the cylindrical face of the hole in the blue part around the red part.

6. Select the Radial Fill check box in the Fill Direction group. This will present the preview shown in Figure 12.7 of a weld closing the gap between the parts.

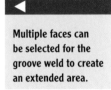

Multiple faces can be selected for the groove weld to create an extended area.

7. Click OK to create the groove weld.

The groove weld could be left as is, or it could be added as part of a manufacturing process that would now add another weld to it.

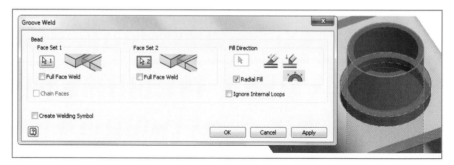

FIGURE 12.7 The preview of the groove weld

Adding a Cosmetic Weld and Weld Symbols

The fillet and groove are the most common welds and therefore generate physical geometry to represent them, but there are many types of welds that need to be documented and considered in the weldment. A cosmetic weld will apply a weld to an assembly but will not build the geometry itself.

1. Make certain that the 2013 Essentials project file is active, and then open c12-05.iam from the Assemblies\Chapter 12 folder.

2. Access the Weld tools by clicking Welds from the marking menu or from the Process panel on the Weld tab.

3. In the Weld panel of the Weld tab or from the marking menu, click the Cosmetic Weld tool.

4. In the cosmetic weld, set the Area value to 6 mm^2.

5. Leave the Extents set to All, and select the Create Welding Symbol check box.
    This expands the dialog box and allows you to select the specific type of weld that the cosmetic weld will represent. The default is a fillet weld for the opposite side.

6. Click the Swap Arrow/Other Symbols icon to make the fillet weld for the Arrow side.

7. Make sure the Bead selection icon is active, and click the edge shown in Figure 12.8.

8. Click OK to place the weld, which is represented by a highlighted edge and the weld symbol shown in Figure 12.9.

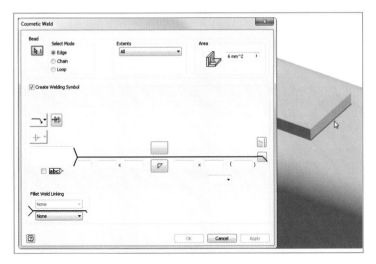

FIGURE 12.8 Select the edge for the cosmetic weld.

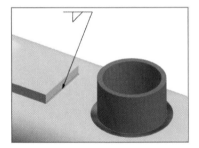

FIGURE 12.9 A weld symbol can be added to any weld placed.

TIP You can set the weld symbol inside the model. Move near the midpoint of the highlighted edge to show the concealed symbol, and use the grip points to relocate it.

Adding an Intermittent Fillet Weld

Sometimes referred to as a *stitch weld*, this feature creates fillet welds of a set length and distance apart, with the ability to specify a limited number.

1. Make certain that the 2013 Essentials project file is active, and then open c12-06.iam from the Assemblies\Chapter 12 folder.

2. Access the weld tools by clicking the Welds icon from the Process panel on the Weld tab or the marking menu.

3. When the tools appear in the Weld panel of the Weld tab, click the Fillet tool.

4. When the Fillet Weld dialog opens, set the first fillet size to 6 and the second to 4.

5. Change the contour from Flat to Convex, and set the offset to 1.

6. Set the top Intermittency value to 9 and the one below it to 10.

7. Click the first bead selection on the side of the block and the second on the top face of the tube, as shown in Figure 12.10.

8. Click OK to place the fillet welds.

FIGURE 12.10 A preview of the intermittent fillet weld

Now you've added the final fillets to this weldment, and what was an assembly of parts can be produced as a single deliverable.

Adding Machined Features to the Weldment

Because of the distortion from welding components together and the need for some features to pass through several of the welded components, sometimes it is best to just machine features after components are welded together.

1. Make certain that the 2013 Essentials project file is active, and then open c12-07.iam from the Assemblies\Chapter 12 folder.

2. Access the machining tools by clicking the Machining icon on the Process panel of the Weld tab or the marking menu.

There is already a sketch in the assembly; it was added in the machining process and appears in the Browser that way.

3. Click the Extrude tool from the Preparation And Machining panel.

4. Use the Extents drop-down in the Extrude dialog to set its value to To.

5. Click the interior surface of the red part for the termination. The preview will appear like Figure 12.11.

6. Click OK to remove all material between the sketch plane and the selected face.

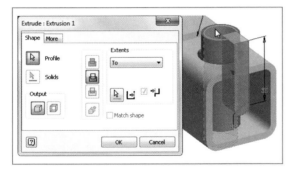

FIGURE 12.11 A preview of the intermittent fillet weld

Like preparation features, the cuts made by the extrusion do not appear in the source files. They remove the original component geometry and the geometry created by the welds themselves.

Documenting Welds and Weldments

Inventor keeps track of the physical properties of the welds as they're added. It also keeps track of the various states of the model so that the preparation, welds, and machining geometry can be documented as a workflow.

Extracting the Physical Properties of the Beads

To help estimate the materials needed to create a weldment, you have the BOM for the components, but you also need the volume and mass of the weld beads.

1. Make certain that the 2013 Essentials project file is active, and then open c12-08.iam from the Assemblies\Chapter 12 folder.

2. Expand the Welds category in the browser, and right-click the Beads folder.

3. Select iProperties from the context menu.

4. Switch to the Physical tab, and click Update to get the mass and volume information on the weld material for the assembly. See Figure 12.12.

    You can also get the properties for individual beads by generating a bead report.

5. Close the iProperties dialog, and click the Bead Report tool from the Weld panel on the Weld tab.

6. When the weld bead report is opened, click Next.

7. Choose a name and destination for the XLS file. Figure 12.13 shows an example.

FIGURE 12.12 The mass and volume of the welds are calculated even for the cosmetic welds.

Document	ID	Type	Length	UoM	Mass	UoM	Area	UoM	Volume	UoM
C:\Inventor 2012 Essentials\Assemblies\Chapter 12\c12-08.iam										
	Fillet Weld 1	Fillet	105.872	mm	0.011	kg	2.12E+03	mm^2	1.41E+03	mm^3
	Fillet Weld 2	Fillet	259.398	mm	0.038	kg	5.59E+03	mm^2	4.90E+03	mm^3
	Fillet Weld 3	Fillet	109.196	mm	0.011	kg	2.04E+03	mm^2	1.34E+03	mm^3
	Fillet Weld 4	Fillet	109.196	mm	0.011	kg	2.04E+03	mm^2	1.34E+03	mm^3
	Groove Weld 1	Groove	N/A		0.017	kg	2.12E+03	mm^2	2.19E+03	mm^3
	Cosmetic Weld 1	Cosmetic	30	mm	0.001	kg	6	mm^2	180	mm^3
	Fillet Weld 5	Fillet	28	mm	0.002	kg	383.879	mm^2	303.85	mm^3
	Fillet Weld 6	Fillet	28	mm	0.002	kg	383.879	mm^2	303.85	mm^3

FIGURE 12.13 The physical properties of each weld

 TIP Even though there is no physical geometry, a cosmetic weld is capable of contributing to the mass calculations if you apply an Area value in the dialog box.

Creating Weldment Drawings

The purpose of a weldment environment is to accommodate the needs of the production workflow. Documenting the steps is also important. Inventor can create drawing views of the stages of the weldment.

1. Make certain that the 2013 Essentials project file is active, and then create a new drawing using the ISO.idw template under the Metric folder under Templates in the New File dialog.

2. Create a base view from the marking menu or the Create panel of the Place Views tab.

3. Click the Open Existing File button, browse to Assemblies\Chapter 12, and click the c12-08.iam file.

4. Set the orientation of the view to Iso Bottom Left and the scale to 1:3.

5. Click the Shaded style.

6. Switch to the Display Options tab, and uncheck Tangent Edges.

7. Change to the Model State tab in the dialog box.

8. In the Weldment group, set the type of view you want to place to Assembly, and then click a location on the drawing for the view, as shown in Figure 12.14.

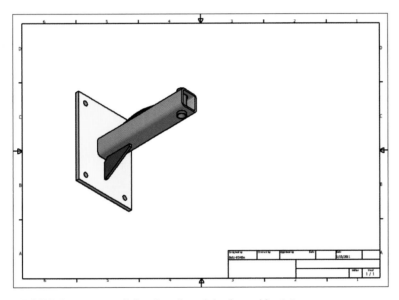

FIGURE 12.14 A drawing view of the Assembly state

9. Start the Base view tools again.

10. Repeat steps 3 through 7.

11. Set the view type for the weldment group to Machining, and place the view so that your drawing looks like Figure 12.15.

The intermediate steps of preparations can be shown in the drawing as well, and any type of drawing can be created from these states, just as you would any other file.

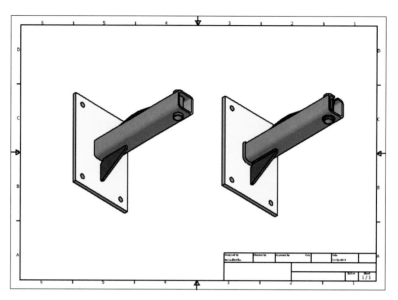

FIGURE 12.15 The drawing showing the beginning and end of the process

THE ESSENTIALS AND BEYOND

Working with tools whose approach is based on the production process builds efficiency into the design from the beginning and makes it easier to edit should production require that changes be made to the weldment.

The separate environments within a weldment make it easier to keep the focus on the steps of manufacturing a welded component and making sure that each step has all the information needed to complete the task.

By offering the flexibility of the cosmetic weld in addition to welds that add physical geometry to the weldment, like fillet and groove, Inventor ensures that you'll be able to document and plan for any type of welding that your design needs.

ADDITIONAL EXERCISES

▶ Try using a Groove weld to connect the gusset to the tube instead of the fillet weld.

▶ See how different load types affect the results of the weld calculator.

▶ Add additional types of features to the machining state.

▶ Add welding symbols to the drawing from the Symbols panel of the Annotate tab.

CHAPTER 13

Creating Images and Animation from Your Design Data

2D drawings have long been the standard in engineering communication, but increasingly they're seen as an incomplete means of communication. Building 3D geometry and using it only as a resource for 2D drawings is selling your work short. There is much more you can do with it.

Let's start with manufacturing and maintenance documentation. The exploded view is a great communication tool. In Chapter 6, "Creating Advanced Drawings and Detailing," you used an Autodesk® Inventor® 2013 software presentation file to create an exploded view drawing. In this chapter, you will see how that presentation file was created.

Another valuable asset that you can create from the 3D work is a rendering or rendered animation. Marketing can then use these files to promote the products while the product is in engineering. You can also use this information to lobby inside the company for the opportunity to develop a concept.

▶ **Developing an exploded view**

▶ **Creating renderings and animations**

▶ **Building options to refine scenes**

▶ **Making a movie of the assembly**

Developing an Exploded View

Certification Objective

The presentation file exists to enable the user to pull the assembly apart without really affecting the assembly. A presentation file is based on an assembly but does not change it. It will, however, reflect any changes made to the assembly just as a drawing view would. You'll use this behavior to create an exploded view of an assembly and then add simple animation.

Using the Automated Technique

When creating the exploded view in the presentation file, you have two primary ways of placing the assembly. One automatically separates the parts based on the values of the assembly constraints; the other simply places a representation of the intact assembly in the Design window for you to separate.

1. Verify that the 2013 Essentials project file is active, and then open the c13-01.iam file from the Assemblies\Chapter 13 folder.

2. Click the New tool on the Quick Access toolbar.

3. Double-click the Standard (mm).ipn template in the Metric folder to create a new presentation file.

Create View

4. Once the new file is loaded and the Ribbon updates with the new Presentation tab, click the Create View tool on the Create panel.

5. Set the Explosion Method option to Automatic, make the Distance value 60 mm, and click OK.

6. Use the Orbit tool to see the results, as shown in Figure 13.1.

FIGURE 13.1 Creating an exploded view with trails

This assembly was constructed using Insert constraints to align the parts. Typically the Automatic explosion method does not produce such ideal results if the assembly isn't built this way.

Building One Step at a Time

Certification Objective

When you want complete control over the explosion separation and order, it is necessary to place the view intact and then move the components using the tools specific to the presentation file. In the following exercise, you will use the most commonly used tool in the presentation file, Tweak Components:

1. Verify that the 2013 Essentials project file is active, and then open the c13-01.ipn file from the Assemblies\Chapter 13 folder.

2. Click the Create View tool in the marking menu.

3. In the Select Assembly dialog box, use the Open Existing File tool to select the c13-01.iam file from the Assemblies/Chapter 13 folder.

   This might seem a little strange, since there is already one exploded view of the same assembly already in the file, but like drawing files, presentation files can document as many assemblies as you like in one file.

4. Make sure the Explosion Method option is set to Manual, and click OK.

5. Click the Tweak Components tool in the Create panel of the Presentation tab or in the marking menu.

   From the number of icons with red arrows, you can tell this dialog box wants a lot of user input.

6. Click the face of the blue part that the washer rests on for the direction.

7. For the components, click the bolt and the washer.

8. You can bypass setting a trail origin and just click in an empty portion of the Design window and drag to the right roughly 60 mm, as shown in Figure 13.2.

TIP While you are creating a tweak or series of tweaks, you can add or remove components from your selection without having to reselect the direction.

9. Click the blue part to add it to the selection, and drag the components to the right again by approximately 60 mm.

FIGURE 13.2 Drag the components apart.

10. In the value box in the Transformations group in the dialog box, type 65, and click the green check mark to set the value of the last tweak to 65 mm.

11. Now hold the Ctrl key and click the blue part and the washer to remove them from the selection.

12. Drag the bolt to the right approximately 50 mm, and click the Clear button in the dialog box.

    This clears your previous selections for Direction and Components but does not close the Tweak tool. This is a great way to change direction or focus without having to restart the tool.

13. If it's not already highlighted, click the selection icon for Direction, and then click the shaft of the bolt.

14. Change the Transformation type from Linear to Rotational by clicking the Rotational radio button.

15. Select the bolt, change the Transformation value to 720, and click the green check mark.

16. Click Close to finish adding tweaks to the exploded view, which will closely resemble Figure 13.1.

As it is, the current view would work well for showing someone how this assembly is built. But why limit yourself to a static view when you can easily build an animation from the work you've already done?

Controlling the Assembly Instructions

Certification
Objective
Everything needed to create an animation is already done. Now, you need to be sure that this animation is done in the correct order and add controls for visualization and modify some of the values. Animations can show assembly or disassembly, and as you will see in the next exercise, you can control the order in which the parts are moved:

1. Verify that the 2013 Essentials project file is active, and then open the `c13-02.ipn` file from the `Assemblies\Chapter 13` folder.

2. Click the Animate tool in the marking menu or on the Create panel of the Presentation tab.

3. When the Animation dialog box opens, click the Play Forward icon.
    The tweaks you added in the previous exercise animate in reverse order as the assembly comes back together.

4. Click Reset to restore the exploded view.

5. Expand the dialog box to see the list of animation sequences shown in Figure 13.3.

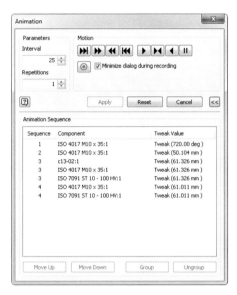

FIGURE 13.3 The list of sequences making up the animation

6. In the Animation Sequence list, click Sequence 1, and click the Move Down button until it is at the bottom of the list.
    This sequence will now be labeled Sequence 4.

7. With Sequence 4 still highlighted, hold the Ctrl key, and click the two members of Sequence 3.

8. Click the Group icon to combine the three tweaks into a new Sequence 3, and click Apply.

9. Click the Auto Reverse button to see that the rotation of the bolt is now combined with the final step in the assembly process and to see the assembly expand again afterward.

You can change the Interval value to speed or slow the sequences, and you can set the number of repetitions for the animation. But there are options that can add additional control; you will use them next.

Adding the Details

The order of the assembly is now correct, and the addition of the bolt turning gives the animation a little flair. To make the animation complete, you can have the sequences appear differently by changing the camera position and part visibility. You can also take this animation and export it as a video file for review by others.

1. Verify that the 2013 Essentials project file is active, and then open the c13-03.ipn file from the Assemblies\Chapter 13 folder.

2. At the top of the Browser is the Browser Filters icon. Click it, and select Sequence View in the context menu.

    The Browser updates and shows a series of tasks as well as the structure of the assembly that acts as a parent to the presentation file.

3. Expand the elements under Task1, and move your cursor over them to see that they highlight the components involved in the sequence.

There are also filters to view the assembly and a listing of each tweak.

4. Expand Sequence 1 so you can see the Hidden folder.

5. At the bottom of the Browser is the assembly. Expand it, drag the c13-01 part under the Hidden folder of Sequence 1, and reposition the assembly to appear like Figure 13.4.

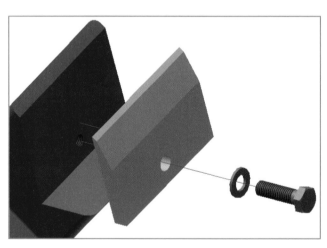

FIGURE 13.4 Each sequence can have a unique camera position to make the actions of the step easier to understand.

6. Double-click the Task1 icon.

Set Camera

7. When the dialog box opens, change the Interval value to 15, click the Set Camera button, and then click OK.

8. Use the Orbit tool to see the top, left, and back of the ViewCube.

9. Double-click the icon next to Sequence 2, choose Set Camera, and click OK.

10. Start the Animate tool again, and click the Record button in the Animation dialog box.

11. When the Save As dialog box opens, enter a filename, set the location to your computer desktop, and click Save.

12. Select Custom Profile from the drop-down. Change the Network Bandwidth value to BroadBand(250Kbps) and the Image Size value to 640 × 480 in the ASF Export Properties dialog box, as shown in Figure 13.5; then click OK.

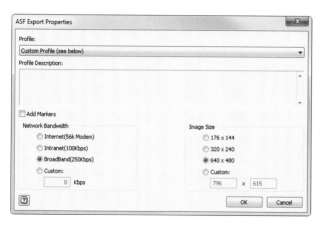

FIGURE 13.5 Select the options for creating the video of the animation.

13. When you return to the Animation dialog box, click the Auto Reverse icon.

14. When the animation is finished, click the Record button again and then the Cancel button to finish the tool.

Go to your computer's desktop, and play the new video you created. Switching the file type to AVI in the Save As dialog box gives you options to create a higher-quality video, but the steps are basically the same. A Completed c13-03 Assembly.wmv video is stored in the Assemblies\Chapter 13 folder.

Creating Renderings and Animations

The presentation file is a great way to quickly communicate with others how your product needs to be built. You can also use the engineering data to interest people in buying your product, but that requires a much higher-quality output. It's time to look at some of the tools of Inventor Studio. These tools give you advanced rendering and animation capabilities and additional types of renderings.

Creating a Still Image

You can render any part or assembly file. The easiest way to do this uses the Realistic visualization mode on the View tab. To have more control, you need to enter the Inventor Studio environment.

1. Verify that the 2013 Essentials project file is active, and then open the c13-02.iam file from the Assemblies\Chapter 13 folder.

2. Make the Environments tab active.

Inventor Studio

3. Click the Inventor Studio icon on the Begin panel to enter the Inventor Studio environment.
   Like the Sketch tab, a contextual Render tab will be highlighted because it is a temporary environment, and you cannot do modeling tasks while in the environment.

Render Image

4. Click the Render Image icon in the Render panel on the Render tab or in the marking menu to open the Render Image dialog box.

5. Without changing any options, click the Render button. After the rendering is finished, it appears like Figure 13.6.

6. Close the Render Output window.

FIGURE 13.6 The initial rendering

7. Change the Render Image dialog box settings to the following:

  ▶ Camera = Essentials 1

  ▶ Lighting Style = Sample 1 Lighting

  ▶ Scene Style = XZ Ground Plane

  ▶ Render Type = Shaded

8. Click Render to create the image shown in Figure 13.7.

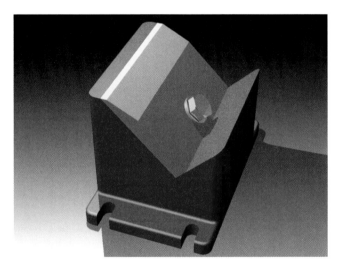

FIGURE 13.7 The rendering with more developed options

Pictorial images are not the only thing needed to show others; sometimes, you also need to create renderings for use in technical illustrations.

9. Close the Render Output window to prepare to make another rendering.

10. In the Render Image dialog box, change the settings to the following:

  ▶ Camera = Essentials 1

  ▶ Lighting Style = (Current Lighting)

  ▶ Scene Style = (Current Background)

  ▶ Render Type = Illustration

11. Select the Output tab, and change the Antialiasing setting to High Antialiasing.

12. Click the Style tab, set Color Fill Source to No Color, and click Render to generate the image in Figure 13.8.

13. Close the Render Image dialog box.

14. Close the assembly file without saving changes.

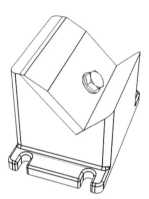

FIGURE 13.8 An illustration
developed after a few settings changes

Even with basic changes to the options, the results are dramatically different. Next, you will see how to make changes that will help you create your own scenes and lighting to add more options.

Building Options to Refine Scenes

To add details and improve the appearance of renderings and animations, you need to expand upon the included library of appearance options, lighting, and scene styles as well as building cameras and local lights.

Customizing Appearance for Rendering

With three color libraries included in Inventor, it is easy to find an option that makes your components look good. Sometimes, you might want something just a little different. You can apply an appearance to a component and then modify that appearance to create a custom look.

1. Verify that the 2013 Essentials project file is active, and then open the c13-02.iam file from the Assemblies\Chapter 13 folder.

2. Click the Tools tab and pick the Appearance tool from the Material and Appearance panel.

3. Click the arrow next to Inventor Material Library in the bottom-left column to display the available appearance options.

4. Click Metal to show the appearance swatches in the right column.

5. Find the Iron – Cast swatch, hover over it, and select the Up arrow icon to add it to the list of Document Appearances at the top, as shown in Figure 13.9.

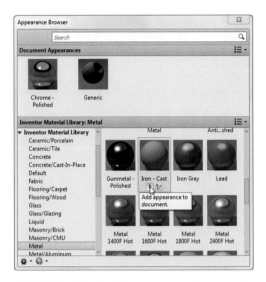

FIGURE 13.9 The Appearance Browser gives you access to appearance options.

6. Locate the icon for Iron – Cast in the Document Appearances list, right-click it, and select Duplicate from the context menu.

7. The new appearance style name is highlighted. Type **Blue - Cast** and press the Enter key to rename it.
   This new appearance option can be edited.

TIP   The Adjust tool in the Material and Appearance panel has great tools for loosely modifying colors. The color ring can set the basic color, and the diamond in the center can be used to adjust hue and tint.

8. Move your cursor over the Blue – Cast appearance and click the eye-dropper icon to open the Appearance Editor dialog box.

9. When the dialog box opens, click the top bar next to the word *Color.*

10. When the Color dialog box opens, select the darkest blue in the top row, and then click OK to make the selection and close the Color dialog box.

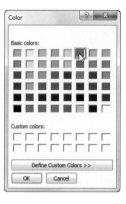

11. In the Bump section of the dialog box, use the slider or enter a value of 50 for the Amount value.

12. Click OK to close the Appearance Editor dialog box.

13. Pan or zoom the Graphics window so the assembly is visible.

14. Click the red base, and then click the Blue – Cast icon in the Appearance Browser dialog box, as shown in Figure 13.10.

Changing the color of a component in the assembly does not change it in the component's file.

15. Close the Appearance Browser dialog box by clicking the Close icon in the upper right.

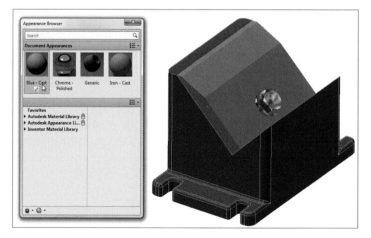

FIGURE 13.10 A component's color can be changed by highlighting it and then clicking a color in the dialog box.

Appearance options can be modified in many ways. By making changes to existing colors or by creating your own, you can easily give your designs a custom appearance.

Changing the Scenery

A compelling visual effect is to place your design in an environment that looks like where it will be located or to create a scene that highlights the design by eliminating the background altogether.

1. Verify that the 2013 Essentials project file is active, and then open the c13-03.iam file from the Assemblies\Chapter 13 folder.

2. Enter the Inventor Studio environment by clicking the Inventor Studio icon on the Environments tab.

3. Click the Scene Styles icon in the marking menu or the Scene panel to open the Scene Styles dialog box.

4. Right-click the XZ Reflective GP style, and click New Scene Style in the context menu.
   This creates a new scene style named Default 1.

5. Right-click the new scene style, and select Rename Scene Style in the context menu.

6. When the Rename Scene Style dialog box appears, give it the name Essentials XZ, and click OK to change the name.

7. Right-click the Essentials XZ style, and click Active in the context menu.

8. On the Background tab, click the Top Color icon in the Colors group, and set the color to black when the Color dialog box appears.

9. Make the Environment tab active, use the slider to make the Show Reflections value 40, and uncheck the Show Shadows box.

10. Click Save and then Done to close the Scene Styles dialog box.

11. Use the Render Image tool to generate an image like that shown in Figure 13.11.

Creating an array of scene styles gives you many options. You can also generate scene styles using images of the environment your designs will be in, including reflections of the lighting for the environment.

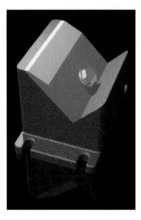

FIGURE 13.11 The rendering
with the modified scene

Applying Different Lighting

Changing the lighting in a rendering can make a huge difference in the quality
of the images you create. Even small changes can have a dramatic effect. For
this example, you will make a simple change to see what the total effect is.

 Lighting Styles

1. Verify that the 2013 Essentials project file is active, and then open the
   c13-04.iam file from the Assemblies\Chapter 13 folder.

2. Click the Inventor Studio icon on the Environments tab.

3. In the Scene panel, click the Lighting Styles tool, or select it from the
   marking menu.

4. Right-click the Sample 1 lighting style, and click Copy Lighting Style.

5. Give the new style the name Essentials 1 in the Copy Lighting Style
   dialog box, and click OK.

6. Right-click the new Essentials 1 style, and click Active in the
   context menu.

7. Expand the style, and click the Blue Fill element.
      The lighting type, placement, direction, intensity, and color can all
   be changed.

8. Click the Illumination tab; then click the Light Color icon, set the
   color to white, and click OK.

9. Use the slider to set the Intensity value to 80, and then click the
   Shadows tab.

10. Select the Sharp Shadows option, and then activate the Directional tab.

11. Set the Longitude and Latitude values to 35, and then click Save and Done to finish the edit.

12. Click the Render Image tool on the Render panel to create a new rendering like Figure 13.12.

FIGURE 13.12 Updating the lighting has a huge effect on the rendering.

Additional shadows and highlights appear in the rendering thanks to the addition of shadows, the repositioning of the light, and the change in its color. Of the things that affect the scene, lighting is the aspect that will take the most practice and produce the most important results.

Adjusting the Camera Settings

The Essentials 1 camera that is used in the previous exercises was created by right-clicking in the Design window and selecting Create Camera From View in the context menu. Now you will modify it to give the renderings a little style:

Certification
Objective

1. Verify that the 2013 Essentials project file is active, and then open the c13-05.iam file from the Assemblies\Chapter 13 folder.

2. Click the Inventor Studio icon on the Environments tab.

3. Expand the Cameras group in the Browser, right-click the Essentials 1 camera, and click Edit in the context menu.

4. In the Camera:Essentials 1 dialog box, change the Zoom value to 36 deg, and set the Roll Angle value to -10.

5. In the Depth of Field group, click the Enabled check box, and set the Near value to 400 mm and the Far value to 430 mm. See Figure 13.13.

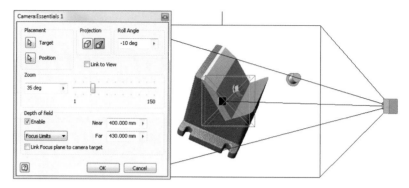

FIGURE 13.13 Editing the camera settings

These changes will rotate the view and zoom in a little closer. The focus limits will put the area between 400 mm and 430 mm in sharp focus, with parts of the model closer or farther from the camera becoming increasingly out of focus.

6. Click OK to make the changes to the camera.

7. Use the Render Image tool on the Render panel to generate the image in Figure 13.14.

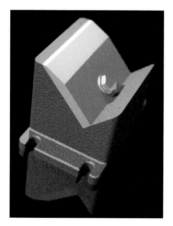

FIGURE 13.14 Setting focus limits in the camera creates blur in portions of the rendering.

Creating cameras is easier done than said, but I still prefer to get a point of view in the Design window and initiate creating a new camera from there. Now, you can leverage all the changes you've made to create an animation of the assembly in action.

Making a Movie of the Assembly

It used to be that making an animation required another application besides the engineering tool. Translating the engineering data to the animation tool was difficult, and the animations often ended up being out-of-date because they couldn't be linked to the latest design changes. Inventor bridges that gap.

Creating the Timeline and Using the Camera

Camera changes in Inventor Studio can be as advanced as having the camera move along a path as it follows another path. But simple changes can add flair to an animation as well.

Certification Objective

1. Verify that the 2013 Essentials project file is active, and then open the c13-06.iam file from the Assemblies\Chapter 13 folder.

2. Open the Environments tab, and click the Inventor Studio icon in the Begin panel.

 TIP Inventor can create multiple animations using different cameras, moving parts, and other variables. To keep track of it all, you have to create an animation within the file.

3. Click the Animation Timeline tool in the Animate panel on the Render tab to create a new animation, and click OK to confirm that Inventor Studio will activate the new animation.

   The Animation Timeline (Figure 13.15) appears across the bottom of the Design window. Toward the right is a drop-down displaying the use of the current view for the animation.

Animation Timeline

FIGURE 13.15 The Animation Timeline is the main resource for editing animations.

4. Click this to make the Essentials 1 camera the active camera for the animation.

5. Just to the right of the drop-down is the Animation Options icon. Click it to open the Animation Options dialog box.

The Video Producer tool in the Animate panel can combine a series of cameras to create a video. These videos can include several types of transition between camera positions.

6. In the dialog box, set the Length value of the animation to 5 seconds, and leave the Default Velocity Profile value set to its defaults, as shown in Figure 13.16; then click OK.

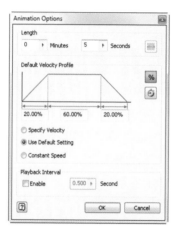

FIGURE 13.16 The Animation
Options dialog box

7. The scale on the timeline has a slider that also displays the time. Click and drag it to read 4.0 (seconds).

8. Click the corner of the ViewCube to display the right, back, and top sides (Figure 13.17), and then click the Add Camera Action icon in the Animation Timeline.

Using Add Camera action repositions the active camera to that point of view at the selected point in time.

FIGURE 13.17 The view that the
animation will have at four seconds

9. Drag the slider toward 0 on the timeline to see how the camera motion is added to the animation.

10. Click the Play tool to see the changes you've made animated.

This is another example of how you can make an advanced tool easy to use. Defining a new path for the camera took only a handful of mouse clicks, and yet the result looks great.

Making the Components Move

Camera control isn't the only thing that has been dramatically simplified in Inventor Studio. The animation can use positional representations to define the beginning and ending conditions for mechanisms in the assembly.

1. Verify that the 2013 Essentials project file is active, and then open the c13-07.iam file from the Assemblies\Chapter 13 folder.

2. Click the Inventor Studio tool in the Begin panel of the Environments tab.

3. Click the Pos Reps tool in the Animate panel of the Render tab.

4. On the Animate tab of the Animate Positional Representation dialog box, leave the Start position set to Master, but change the End position to Open using the drop-down menu.

5. In the Time group, click Specify As Measurement, and set the Start time at 0.2 s and the End time to 1.5 s, as shown in Figure 13.18.

6. Click OK to create the new animation element.

7. Now click the Expand Action Editor icon to see a full view of the timeline and activities on it.

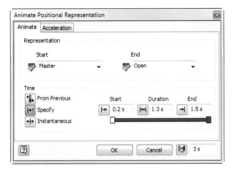

FIGURE 13.18 Setting a transition time to move assembly components

In the timeline, you'll see blue bars that show the duration of the actions on the rows to which they align.

8. Right-click the action on the slider bar with the PRAnim1 (Master<>Open) action, and click Mirror in the context menu, as shown in Figure 13.19.

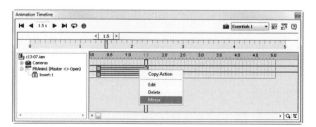

FIGURE 13.19 Events in the timeline can be copied or mirrored rather than defining new ones.

9. When a new bar appears next to the old, drag it all the way to the right so that the reverse action happens at the very end of the animation.

10. Click the Expand Action Editor icon again to reduce the Animation Timeline, and then click the Go To Start icon.

11. Use the Play Animation tool to see the combination of the camera transition and the moving positional representation.

The animation is now quickly taking shape, so you need to add only one more finishing touch to complete the work. The bolt needs to turn to open and close the vise.

Animating an Assembly Relationship

Certification Objective

Assembly constraints reflect how components relate to each other in the real world. It makes sense that they would be used as the foundation of an animation to show how the assembly works.

1. Verify that the 2013 Essentials project file is active, and then open the c13-08.iam file from the Assemblies\Chapter 13 folder.

2. Click the Inventor Studio icon in the Begin panel of the Environments tab to open the Inventor Studio tools.

3. When the Render tab appears, click the Constraints tool in the Animate panel.

The Animate Constraints dialog box has a Select icon depressed and ready to click an assembly constraint from the Browser.

4. Expand the Browser view of the ISO 4017 M10 × 35:1 component, and click the Bolt Angle constraint.

5. In the dialog box, set the End value to 360 deg, click the Specify Time option, and start the action at 0.2 s and end it at 1.5 s, as shown in Figure 13.20.

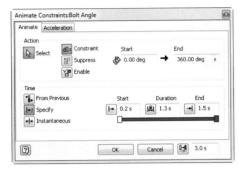

FIGURE 13.20 Animating an assembly constraint

6. Click OK to create the new action.

7. Click the Expand Action Editor icon to see a full view of the timeline and activities on it.

8. Right-click and mirror the action for the Bolt Angle constraint.

9. Drag the new action to the end of the toolbar.

10. Play the animation in the Design window.

Now that you've created all the steps, you need to create the animation of the assembly. Next, you will render the animation to a video file.

Rendering the Animation

The camera moves, the vise opens and closes, and even the bolt turns out and then back in. Now, you need to share your work.

1. Verify that the 2013 Essentials project file is active, and then open c13-09.iam file from the Assemblies\Chapter 13 folder.

2. Switch the Ribbon to the Environments tab, and click the Inventor Studio tool in the Begin panel.

Render
Animation

3. Click the Render Animation tool in the marking menu or in the Render panel of the Render tab to open the Render Animation dialog box.

4. Switch to the Output tab, and change the end of the time range to 5.0 s.

5. If your computer has limited graphics, you can use the drop-down to set the frame rate to 10.

6. Click Render, and the Save dialog box will open. Enter a filename, set the location to your computer Desktop, and click Save.

▶

You can also create an animation as an illustration or as a series of still images.

7. Change the Network Bandwidth value to BroadBand(250Kbps) and the Image Size value to 640 × 480 in the ASF Export Properties dialog box, and click OK.

The animation will be created by generating a rendering of each frame of the animation. The corners of squares will appear one for each processor in the computer. When it is complete, review the finished WMV file in the Completed c13-09.wmv file in the Assemblies/Chapter 13 folder.

THE ESSENTIALS AND BEYOND

Any reuse of the design data for marketing or production documentation, or even just for explaining an idea, multiplies the value of your work. Having drawings, presentation files, and animations all reflect any subsequent changes to that data is further reward for your efforts in creating the 3D geometry.

Creating renderings and animations is easy. Using assembly constraints and maintaining a focus on the function of the assembly will guide you to creating effective animations. When creating animations of presentation files, it is good practice to review the process with people who will produce the design to make sure it is valid.

ADDITIONAL EXERCISES

▶ Consider creating an animation based on a presentation file for shop assembly instructions.

▶ Use the Save Copy As tool to make screen captures of exploded views.

▶ Change lighting types to see the differences and their effect on the rendering.

▶ Save a rendered animation as an AVI file to compare quality and file size.

CHAPTER 14

Working with Non–Autodesk Inventor Data

The Autodesk® Inventor® 2013 software is used in many industries to fulfill several roles. One common role requires users to design tooling and provide other manufacturing support. You might find yourself working with data that was sourced from other programs and that may or may not be 3D. In turn, you might have to give data to users of other systems inside or outside the company.

Architects use the products of mechanical engineers and designers everywhere in the buildings they construct. Heating and cooling equipment, lighting fixtures, doors, windows, and curtain wall are all manufactured and ideal components to be designed using Autodesk Inventor. Versions of Autodesk® Revit® MEP software can connect equipment that requires electricity, requires water, or does air handling with intelligence. If you're a building equipment manufacturer, you can use Inventor to share data with Revit MEP users so they can include your products in their designs.

▶ **Exploring the data formats for Inventor Import and Export**

▶ **Working with AutoCAD data**

▶ **Exchanging 3D data**

▶ **Creating content for building information modeling (BIM)**

Exploring the Data Formats for Inventor Import and Export

By knowing the formats you can work with, you can better communicate with others about how you can most easily work with them. Table 14.1 shows the file formats that Inventor systems can read or write primarily using

Open ➤ Save Copy As. Some of these formats are read or written in less common tools, such as when inserting an image into a sketch.

TABLE 14.1   External data formats Inventor systems can import or export

Import	Export	Format	Purpose
✔	✔	.idw	Inventor 2D drawing file format
✔	✔	.dwg	Inventor, Autodesk® Inventor® Fusion, and Autodesk® AutoCAD® drawing file format
✔	✔	.ipt	Inventor part file format
✔	✔	.iam	Inventor assembly file format
✔	✔	.ipn	Inventor presentation file
✔	✔	.ide	Inventor library feature format
✔	✔	.dwf, .dwfx	Autodesk® Design Review Markup file
✔	✔	.dxf	Autodesk 2D exchange file format
✔	✔	.sat	ACIS kernel-based exchange format
✔	✔	.igs, .ige, .iges	Product-neutral 3D surface file formats
✔	✔	.stp, .ste, .step	Product-neutral 3D solid file formats
✔	✔	.x_b, .x_t	Parasolid binary and text-based exchange formats
✔	✔	.prt, .asm	Creo Elements/Pro (formerly Pro/ENGINEER) part and assembly file formats
✔	✔	.g, .neu	Creo Elements/Pro (formerly Pro/ENGINEER) kernel-based exchange format
✔	✔	.prt, .sldprt, .asm, .sldasm	SolidWorks part and assembly formats

TABLE 14.1 *(Continued)*

Import	Export	Format	Purpose
✔	✔	.prt	Seimens/PLM NX part and assembly formats
✔	✔	.jt	Compressed format file used by Siemens/PLM data management applications
✔	✔	.png	Adobe Portable Network Graphics file
✔	✔	.bmp	Windows raster file
✔	✔	.gif	Graphic information file
✔	✔	.jpg	Common graphic file (Joint Photographic Experts Group)
✔	✔	.tif, .tiff	Tagged image file
✔	✔	.stl, .stla, .stlb	Stereolithography 3D-neutral format
✔	✔	.cgr, .CATPart, .CATProduct	Dassault Catia V5 files
✔		.model, .session, .exp, .dlv3	Dassault Catia V4 files
✔		.fls, .fws, .las, .ptg, .pts, .ptx, .xyb	Point cloud data
✔		.3dm	Rhino data format
✔		.wire	Autodesk Alias file
	✔	.adsk	Autodesk BIM exchange format
	✔	.CATPart	Dassault Catia V5 file
	✔	.pdf	Adobe Acrobat file
	✔	.xgl, .zgl	Product-neutral 3D graphic–viewing format

Working with AutoCAD Data

With millions of AutoCAD users, the chances are excellent that, at some time, you will need to exchange data with an AutoCAD user. Working side by side with AutoCAD users, you can take advantage of the ability to create drawings in DWG format that AutoCAD users can open and to which they can even add data. In Chapter 10, "Working with Sheet Metal Parts," you used AutoCAD data to develop a folded sheet metal part from a flat pattern by placing the AutoCAD data into a sketch.

Opening AutoCAD Data

Often, you only need to view the data to get information or create a print. You can open an AutoCAD drawing file using Inventor without modifying the source file.

1. Make certain that the 2013 Essentials project file is active, and then open c14-01.dwg from the Drawings\Chapter 14 folder.

   Once the drawing opens, a Review tab on the Ribbon (Figure 14.1) provides tools to view, measure, and print the drawing.

2. Zoom in on the left view of the drawing.

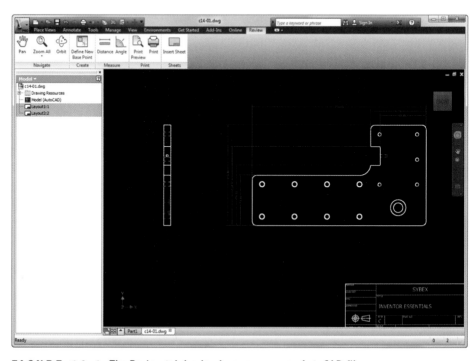

FIGURE 14.1 The Review tab loads when you open an AutoCAD file.

3. Select the Distance tool in the Measure panel to verify the thickness of the material. (The material is 10 mm thick but might be presented in inches.)

Distance

4. Zoom out to see the entire drawing, and then drag a selection window around the geometry of the front view.

5. Press Ctrl+C to copy the geometry to the Windows clipboard.

6. Create a new part file using the Standard (in).ipt template in the English folder, create a new 2D Sketch on the XY plane, and then paste (Ctrl+V) the entities into the sketch by clicking in the Design window to place the data, as shown in Figure 14.2.

7. Close the c14-01.dwg file.

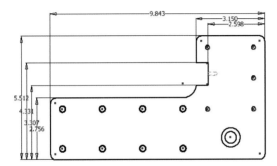

FIGURE 14.2 The dimension values from the AutoCAD data are updated in the new file.

Once the data is placed into a sketch, you can use it to create features using the geometry. You can also use the Inventor drawing tools to add drawing views to the AutoCAD drawing that can be seen by AutoCAD users who open the file.

Importing AutoCAD Data

At other times, you might want to import an AutoCAD drawing to be edited in the Inventor 2D environment.

1. Make certain that the 2013 Essentials project file is active and that there are no files open, and then start the Open file tool.

2. Select the c14-01.dwg file from the Drawings\Chapter 14 folder, but do not open it.

3. Click the Options icon to open the File Open Options dialog box.

Options...

4. Change the option for dealing with a non-Inventor DWG file to Import, and click OK.

5. Now click Open to begin the translation steps.

6. When the DWG/DXF Wizard appears, click Next.

7. In the next dialog box, select the geometry to be imported. In this case, you want the entire drawing, so make sure the All check box is selected in the selection area of the dialog box, and then click Next.

8. In the Import Destination Options dialog box (Figure 14.3), make sure the Constrain End Points option is selected and that the Destination For 2D Data value has New Drawing selected.

FIGURE 14.3 There are many options when converting AutoCAD data into Inventor data.

Selecting Mapping Options in the dialog box allows you to sort AutoCAD layers into separate Inventor sketches. You can also choose to put all geometry into one sketch.

9. In the Templates group, click the icon to browse for a drawing template.

10. Double-click the ISO.idw template on the Metric tab.

11. Click Finish when you return to the Import Destination Options dialog box.

This will open the drawing data in an Inventor drawing. This data is re-created with very good fidelity to the original data, as shown in Figure 14.4.

12. In the Browser, expand ImportedDraftView under Sheet:1 to see the sketches created from the AutoCAD layers.

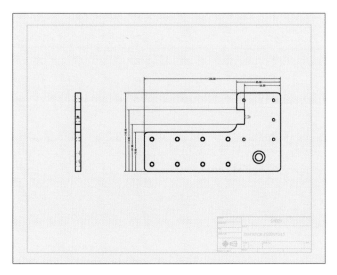

FIGURE 14.4 The dimension values from the AutoCAD data are updated in the new file.

Because the geometry is in sketches, you can change it if you need to make a quick change to a drawing; however, most users want instead to take advantage of the AutoCAD software that comes in the suites that typically accompany Inventor packages.

Editing AutoCAD-Sourced Data

If you need to change the size or content of the data placed in a sketch directly from an AutoCAD file, additional editing might be needed. Editing the parametric data may not be the only change needed. Inventor processing adds some geometric constraints, but others might need to be added because AutoCAD does not understand how to maintain relationships like tangency.

1. Make certain that the 2013 Essentials project file is active, and then open c14-01.ipt from the Parts\Chapter 14 folder.

2. Double-click the 9.843 dimension, and change its value to 10.25.
   As you can see in Figure 14.5, the dimension change made the sketch longer, but it also distorted fillets that were not properly constrained.

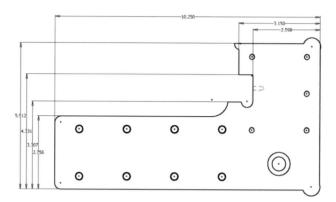

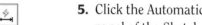

FIGURE 14.5 Changing an underconstrained sketch can cause errors.

3. Press Ctrl+Z or click Undo in the Quick Access toolbar to return the sketch to its original shape.

4. Double-click Sketch1 in the Browser to activate the sketch.

5. Click the Automatic Dimension and Constraints tool in the Constrain panel of the Sketch tab.

6. In the Auto Dimension dialog box, deselect the Dimensions option.
   In this case, you only want the geometric constraints of the sketch evaluated. The tool is capable of adding missing dimensional constraints, as well.

7. Drag a window around the main view of the sketch, and click Apply and then Done to close the dialog box.

8. Change the 9.483 dimension to 10.25 again and see the results in Figure 14.6.

In a more complex sketch, you can use the ability to select specific data to more accurately refine the sketch constraints. Now, we will look at working with 3D data from sources other than Inventor.

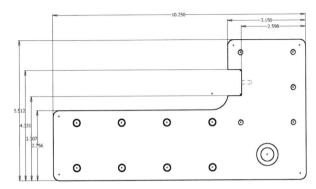

FIGURE 14.6 Adding constraints improves the quality of edits made to the sketch.

Exchanging 3D Data

As you saw in Table 14.1, you can import all of the most common 3D formats (both product-neutral and product-specific) into Inventor. Much of the time, this allows you access to other people's design data with complete accuracy. The original features and data parameters are not accessible, though. Imported data is brought into a single entity referred to as a *base solid*.

Opening assemblies imports the individual components; although they will not be constrained together, they will be positioned as they were in the original data, so the imported data can be used without having to reconstruct the constraints.

Opening Neutral 3D Data

For this exercise in opening neutral 3D data, you will work with a single part, but all the steps are the same for the largest data.

1. Make certain that the 2013 Essentials project file is active, and then start the Open file tool.

2. Use the Files Of Type drop-down to click the STEP Files (*.stp, *.ste, *.step) format, and navigate to the Parts/Chapter 14 folder; then select the c14-01.stp file and open it, as shown in Figure 14.7.

Once the file is opened, it can be saved and used as a resource for creating drawings, assemblies, and any other data. The imported data can have parametric features added, but you cannot edit the features that are on the part.

Use the same process for importing data from other 3D CAD systems.

FIGURE 14.7 The model after import into Inventor

Prior to Inventor Fusion, you could do basic solid editing using tools built into Inventor to edit base solids. This option is still available through the application options.

Editing Imported Data

To edit the base solid, you will use an application that was installed with Inventor, called Inventor Fusion. Inventor Fusion offers many options for editing base solids.

1. Click the Open file tool, and if need be, switch the Files Of Type value back to Autodesk Inventor Files.

2. Make certain that the 2013 Essentials project file is active, and then open c14-02.ipt from the Parts\Chapter 14 folder.

c14-01

3. Locate the imported data icon, which includes the name of the imported file, in the Browser; then expand it, right-click the Base1 element, and select Edit Solid in the context menu.

When Inventor Fusion is used to edit an Inventor part, the changes can be saved to a DWG file; therefore, an Inventor user can review the changes before adopting them.

4. When the instructions for returning to the Inventor appear, read them and then click OK.

5. The Inventor session will be minimized, and Inventor Fusion will open with the model geometry in it.

   In the next two exercises, you will use several features that are unique to Inventor Fusion.

   For the most part, the Inventor Fusion interface is similar to Inventor. There's a Ribbon, a Browser, a ViewCube, and a Navigation Bar.

6. Start the Find Features tool from the Manage panel of the Home tab.

   Starting the Find Features tool opens a contextual tab on the Ribbon where you select which features you want to find in the selected geometry.

7. Deselect the Select All option in the Feature Types panel. This clears all the options.

8. Select the Fillet option, and then click and drag a selection window around the entire part in the Graphics window or pick the icon for the Solid model from the Browser.

9. Once the entire model highlights, click OK in the Feature Recognition panel or right-click in an open area of the Graphics window and pick OK from the compass of the marking menu.

   Several fillets will appear in the Browser. You can use these icons to select fillets. Once a fillet is selected, you can select a feature to edit or remove from the model.

10. While holding the Ctrl key, pick each of the fillets in the Browser, as shown in Figure 14.8.

11. Once they're selected, right-click and click Delete in the marking menu. You can select the fillets in multiple steps if needed.

FIGURE 14.8 Select all the fillets on the part.

12. Once the fillets are gone, the model should have only sharp edges. Before you make further edits, let's change the unit of measurement.

13. In the lower-right corner, hover over the icon that shows the current units (cm), and click mm from the drop-down menu (Figure 14.9).

Now that you've simplified your model, you can begin to change its size and shape. Inventor Fusion has several options for moving faces on a par. You can add features as well as edit them, and the changes you make will be carried back into Inventor when you are finished.

The Simplify tool in the Simulation panel can be used to select or even delete features from the Browser based on their size.

FIGURE 14.9 Set the unit of measure to mm.

Let's make some changes to the model, including a few that can't be done this way from within Inventor.

1. Right-click in the Inventor Fusion Design window, and select Move in the marking menu.

2. Click the top of the handle, and an icon called a *triad* will appear as the face highlights. Drag the yellow arrow down 5 mm or enter the value -5 mm to move the face, as shown in Figure 14.10.

    Zooming in closer to the model reduces the distance between snapping points. Try zooming in if the face tries to move more than 5 mm.

3. Right-click in an open area of the Graphics window and either click OK or drag to the right to complete the change.

    In the Move tool, dragging the arcs on the glyph can change the angle of a face as well. Now that you've thinned the top of the part, let's add contour to the underside.

4. Click the Assign Symmetry icon in the Form Edit panel.

5. When three planes appear through the center of the part, click the longest, vertical plane shown in Figure 14.11, and then click OK to set the value.

6. Start the Edit Edge tool from the Form Edit panel, and click the edge shown in Figure 14.12; then move your cursor near the middle edge, and click to place a grip point.

7. Click the front face on the ViewCube to rotate the view.

    Setting this view will isolate the movements of the point to be parallel to the view. This can be a very effective way to control the edits made to the model.

8. Click the point in the middle of the edge and drag it downward, as shown in Figure 14.13. Release the mouse button, and click OK to finish the edit.

You can select a plane to isolate movement as well, or you can freely move a point in space.

9. Click the Return To Autodesk Inventor icon in the Inventor panel to update the model in Inventor with the changes made in Inventor Fusion.

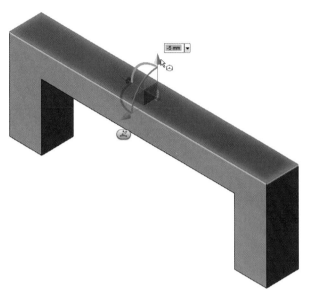

FIGURE 14.10 You can drag to edit the position of the face.

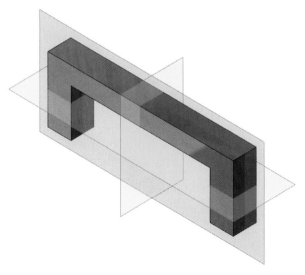

FIGURE 14.11 Clicking a symmetry plane makes edits appear on both sides of the plane.

FIGURE 14.12 Adding an editing point to an edge

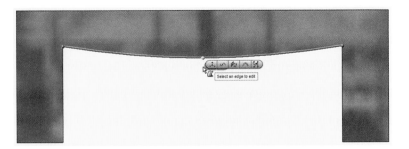

FIGURE 14.13 Dragging a point on an edge can reshape the part.

You can use the Edit Edge tool to make very radical geometry changes to a model. These changes can be made to a parametric Inventor model as well, but it is best to make the changes before finishing touches such as fillets are added.

Exporting Data for Rapid Prototyping

Autodesk's philosophy of digital prototyping calls for the designer to develop and evaluate the function of a design in the digital world to reduce the number of physical prototypes needed. A tool that has been gaining popularity is the use of rapid prototyping equipment, which, for design reviews, provides the tactile feedback that can't be done in the digital domain. These systems will accept a large number of formats, but the one that was created specifically for this process is the stereolithography (STL) model.

1. Make certain that the 2013 Essentials project file is active, and then open c14-03.ipt from the Parts\Chapter 14 folder.

2. Open the Application menu, expand the Save As menu, and click the Save Copy As tool.

3. In the Save Copy As dialog box, change the Save As Type value to STL Files (*.stl), and then click the Preview button.

   When the preview window opens, the model will appear somewhat faceted. You can vary the amount of smoothness in order to judge whether a part that can be created more quickly will give you the level of quality needed.

4. Click the Options button at the top of the window.

5. When the STL File Save As Options dialog box opens, as shown in Figure 14.14, set Surface Deviation to .01, and click OK.

   Check the minimum resolution of your rapid prototyping equipment to avoid creating models whose surface quality cannot be met.

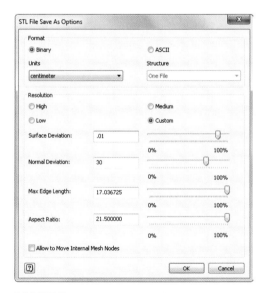

F I G U R E   1 4 . 1 4   Editing STL file export options

6. Click the Show Facet Edges icon to display the edges of the model facets, as shown in Figure 14.15.

7. Click Close to return to the Save Copy As dialog box, set a path, and click Save to create the file.

Having additional control over the model saves time and money and makes the rapid prototyping process more valuable.

FIGURE 14.15 The edges of the model can be displayed to help you make better decisions about resolution.

Creating Content for Building Information Modeling

Building information modeling (BIM) is a philosophy in the architecture, construction, and engineering realm that is very similar to digital prototyping. Creating a database with as much detail about the elements and structure of the building as possible enables it to be built more efficiently, and the model can be used for maintenance in the future.

Autodesk® Revit® is used to design the building's mechanical, electrical, and plumbing components. Much of the equipment used in these trades is created by people who can design it in Inventor. If these vendors can make it easy to put their equipment in a building design, they'll have a competitive advantage. Inventor enables these companies to generate this content for their potential customers.

Simplifying and Securing Your Design

Sharing your design with potential customers is great as long as you don't put your intellectual property at risk. You can use the Shrinkwrap tool to create simplified models any time you like. It allows you to build substitution level-of-detail representations.

In this exercise, you'll build a simplified part to which to add BIM information:

1. Make certain that the 2013 Essentials project file is active, and then open c14-01.iam from the Assemblies\Chapter 14 folder.

2. Click the BIM Exchange tool in the Begin panel of the Environments tab. This will launch the BIM Exchange tab on the Ribbon.

BIM
Exchange

3. Click the Shrinkwrap Substitute tool from the Component panel.

Shrinkwrap
Substitute

4. You can define filenames and paths in the New Derived Substitute Part dialog box, but for this exercise, click OK to accept the defaults shown in Figure 14.16.

FIGURE 14.16 Set the filename and path information in the New Derived Substitute dialog box.

5. In the Assembly Shrinkwrap Options that appear next, click the Range (Perimeter) radio button, and set Min to 5 mm and Max to 1000 mm.

6. For the Simplification mode, select Remove Geometry By Visibility, and leave the slider at 0 percent.

7. Set the style to Single Solid Body Merging Out Seams Between Planar Faces, and then click the Preview button. Figure 14.17 shows the results.

8. Click OK to generate the file using the options.

9. When the simplified model becomes the active level of detail for the assembly, notice that the vent in the top of the housing and the openings on the hose fittings are capped to remove geometry.

10. Switch your Ribbon to the View tab, and use the Section View drop-down menu to click the Half Section View tool.

11. Click the top of the model, and drag the sectioning plane halfway into the part to see that the two internal components have been removed. Your model should match Figure 14.18.

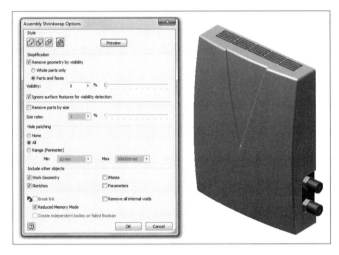

FIGURE 14.17 The Shrinkwrap model can be created as a single or multibody part.

FIGURE 14.18 Simplification can close gaps and remove components from the assembly.

12. Click the check mark in the dialog box showing the distance of the cutting plane.

You can use the Section View tool to expose internal detail and continue to model with it switched on.

13. Using the Section View drop-down menu, click End Section View to restore the view of the assembly.

With the model simplified, it is now ready to have connection information added to it that will ensure your product will be properly connected in the BIM model.

Authoring MEP Content

In this exercise, you'll build a simplified part to which to add BIM information:

1. Make certain that the 2013 Essentials project file is active, and then open c14-02.iam from the Assemblies\Chapter 14 folder.
   The assembly opens with the substitute level of detail active.

2. Start the BIM Exchange tool in the Begin panel on the Environments tab.

3. Click the Pipe Connector tool on the BIM Exchange tab's MEP Author panel.

Pipe Connector

4. When the Pipe Connector dialog box opens, click the round face on the bottom fitting, and click the Make Direction icon so the flow is going outward.

5. The diameter is filled in automatically. Set the Nominal_Diameter value to 3/8 in.

6. Set the System Type drop-down to Domestic Hot Water, as shown in Figure 14.19.

7. Click Apply to add the connector to the part.
   After you place the connector, you can use the dialog box to place another.

8. Click the round face of the upper fitting, and use the direction flip button to show the flow (green arrow) going into the model.

9. Set Nominal_Diameter to 3/8 in, but set System Type to Domestic Cold Water; then click OK.

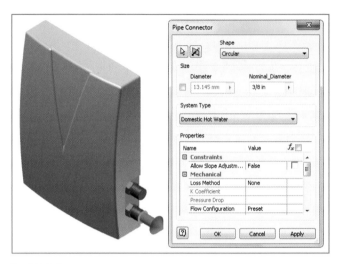

FIGURE 14.19 Adding the hot water output to the heater

10. Rotate the model to see the top, left, and front faces of the ViewCube.

Electrical
Connector

11. Start the Electrical Connector tool from the MEP Author panel.

12. Click the curved edge on the side of the model opposite the water pipe connectors.

13. Set System Type to Power Balanced, change Number Of Poles to 3 using the drop-down as shown in Figure 14.20, and click OK.

14. Click the Export Building Components tool on the Manage panel.

Export Building
Components

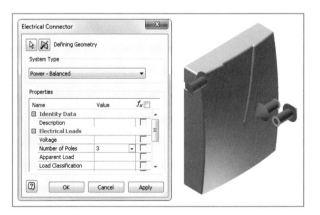

FIGURE 14.20 Adding the electrical connection to the product

15. Click the Browse button under the component type.

16. In the Classification dialog box (Figure 14.21), navigate to the 23.65.35.11.14 Instantaneous Water Heater, and click OK.

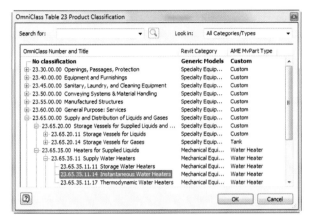

FIGURE 14.21 Loading the appropriate product definition so the Revit MEP software can use your data properly

In the Export Building Components dialog box, you can add the information on your product including a URL for support or installation documentation. This gives the customer access to your data at any time in the life cycle of the building.

17. Click OK, define where to place the new file, and then click Save.

Communication with systems other than Inventor and providing rich data for people using your products give you flexibility to be more competitive and produce higher-quality designs. This is a great combination.

THE ESSENTIALS AND BEYOND

It is very likely you will need to share data with people using systems other than Inventor. Inventor can import data from nearly any system and can export data in a number of formats accepted by other systems.

AutoCAD has millions of users (most likely including you), and you have several ways to edit imported data. Using Inventor Fusion can give old data created in other systems new life without having to start from scratch. Inventor Fusion can also be used as a conceptual modeler to quickly develop new ideas before using Inventor to prepare them for production.

(Continues)

THE ESSENTIALS AND BEYOND *(Continued)*

The BIM content development is priceless if you make equipment that goes into buildings. Getting your equipment included in the design because you've made it convenient to be included can give you a huge advantage over competitors using systems other than Inventor.

ADDITIONAL EXERCISES

▶ Ask a vendor or customer for a noncritical model to test the import and export capabilities of Inventor with them.

▶ Print an existing AutoCAD drawing to see whether the quality is comparable.

▶ Use Inventor Fusion to remove the holes in the bottom of the imported file.

▶ Create a model of an appliance in your home and categorize it using the MEP export tools.

CHAPTER 15

Automating the Design Process and Table-Driven Design

No matter how quickly or easily you can build parts and assemblies, the repetition of building designs that are essentially similar is still a drain on resources. Certain parts or products can easily be defined by tables or dimensions of rules. The Autodesk® Inventor® software works with these types of design on a few different levels.

Using iParts and iAssemblies, it's easy to create a family of parts using a table in the part or an external spreadsheet to manage the variations on the size and features of the components. Using iAssemblies, you can configure variations on an assembly in much the same way, and the process remains manageable.

The next level uses a technology called iLogic to establish rules that define what a valid part is and how parts are allowed to be combined. With the rules established, a request can be made to develop a unique part or assembly, and as long as it meets the criteria of the rules, it can be developed by the system with minimal input from the user.

▶ **Building a table-driven product**

▶ **Expanding the control options**

Building a Table-Driven Product

A familiarity with Inventor's ability to share parameters is important to working with configurations of parts and assemblies, but it's not as important as being familiar with your product. Knowing what does and doesn't change in your product will help you narrow down the number of variables. It can also help you decide where the values will be stored. Like sharing a sketch that can define many features, setting up a part that contains many or all of the parameters can make building associations easier.

Creating a Named Parameter

Every dimension or feature value in Inventor is given a parameter name like d0, d1, or d13405, but they can be hard to keep track of. In Chapter 3, "Learning the Essentials of Part Modeling," you learned that typing **Width=100** would rename the d0 dimension to Width while creating it. You can also edit these values after the fact using the Parameters table. Later, when you create a tabled part, Inventor searches for dimensions to which you gave special names and automatically includes them in the table.

1. Make certain that the 2013 Essentials project file is active, and then open c15-01.ipt from the Parts\Chapter 15 folder.

2. Expand the Drawer Face feature in the Browser, and edit Sketch1, which was consumed by the drawer face.

3. Locate the Dimension Display tool on the status bar. Click it and switch the display to the Expression style.

    This updates the sketch dimension to display the name and current values, as shown in Figure 15.1. Note that the width is controlled by d0 and the height by d1.

FIGURE 15.1 Dimension Display can change the appearance of sketch dimensions.

Parameter names are case-sensitive and cannot contain spaces or use some special characters.

4. Use Finish Sketch to go back to the solid model.

5. Click the Parameters tool on the Quick Access toolbar or the Parameters panel of the Manage tab.

6. In the Parameters dialog box, click the cell under Model Parameters for d0.

7. Type **width** for the new parameter name; then click the cell in the Equation column that shows the current value of 18 in and change it to 12 in.

8. Click the cell for d1, and enter height for its new name; then change its Equation value to 3 in, as shown in Figure 15.2.

9. Click Done to close the dialog box and return to the part.

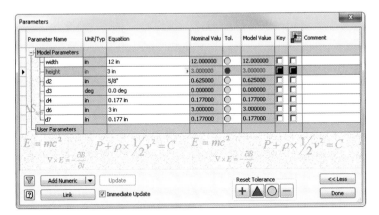

FIGURE 15.2 Editing parameter names and values through the dialog box instead of the sketch

The model will update, showing the changes to the values of the dimensions on the part. Giving a parameter a meaningful name makes it easy to use the Parameters dialog box to edit the part.

Linking Parameters to an External Source

A common technique for building a table-driven product is to use a Microsoft Excel spreadsheet to control the model parameters. Connecting to the parameters is easy, as you'll see in this exercise:

1. Make certain that the 2013 Essentials project file is active, and then open c15-02.ipt from the Parts\Chapter 15 folder.

2. Start the Parameters tool, and in the dialog box, click the Link button.

3. In the Open dialog box, double-click the c15-02.xls file in the Parts\Chapter 15 folder.
   The parameters and their values from the spreadsheet appear under the file path.

4. In the Equation value area for d0, right-click and click List Parameters in the context menu.

5. When a small dialog box listing the two parameters appears, click the Width parameter.

6. Type **height** for the value of d1, and click Done to update the model. See Figure 15.3.

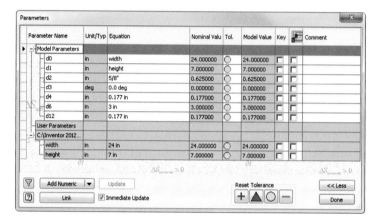

FIGURE 15.3 Parameter values can be linked to cells of a spreadsheet.

7. Double-click the mouse wheel to zoom all to see the part.

8. Open the c15-02.xls file from the Parts\Chapter 15 folder in Excel, and change the value for Width to 12. Close the spreadsheet, saving the change.

9. In Inventor, click the Local Update icon in the Quick Access toolbar to refresh the model.

Another technique that can save a lot of trouble is to link component parameters together. This avoids the need to edit two components to keep them fitting together or to use a spreadsheet to combine parameters.

1. Make certain that the 2013 Essentials project file is active, and then open c15-01.iam from the Assemblies\Chapter 15 folder.

2. Click the Constrain tool in the Position panel of the Assemble tab or the marking menu.

Constrain

3. In the Place Constraint dialog box, click the Constraint Set tab.
 In the Design window (Figure 15.4), you can see two UCS triads. These can be placed apart from the body and be used to align parts

rather than using constraints between faces and edges or traditional constraints.

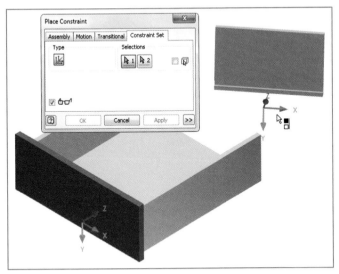

FIGURE 15.4 UCS features can be placed in the components and used to assemble them.

4. Click the two UCS triads in the assembly to position the two components, and then click OK to complete the constraint.

The components do not appear to be aligned correctly, because the position of the UCS in the part that makes up the back needs to be changed along with the height and width of that part. The parameters that control this are embedded in the front of the drawer. Now, you will connect the back to the parameters of the front to make them consistent.

5. Change your Ribbon to the View tab, and click Object Visibility in the Visibility panel to extend its drop-down menu.

6. Uncheck UCS Triad and the other UCS elements to remove them from the Design window.

7. Double-click the c15-03 part in the Browser or in the Design window to activate it.

8. Open the Parameters dialog box from the Quick Access toolbar.

9. Click the Link icon, and when the Open dialog box comes up, change the Files Of Type drop-down to Inventor Files.

10. Double-click the Parts\Chapter 15\c15.04.ipt file.

    When the Link Parameters dialog box opens, colored balloons indicate the linked (yellow) or unlinked (gray) status of each user-named and special parameter. Clicking a gray balloon turns it yellow and links that parameter into your active part.

11. Deselect the Show All Parameters check box, and then click the drawer_depth, drawer_side_height, back_width, and bottom_thickness parameters (Figure 15.5). Click OK to close the dialog box.

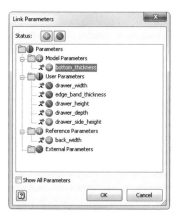

FIGURE 15.5 Selecting parameters to link into the active part

12. Equate the named parameters in the part to the linked parameters as listed:

    ▶ width = back_width

    ▶ height = drawer_side_height

    ▶ thickness = bottom_thickness

    ▶ offset = drawer_depth

13. Click Done to update the part, and then click the Return icon on the Ribbon to go back to the assembly.

> **Some users create the shared parameters in the assembly itself or in a part file with no geometry.**

The back will move into place, sized based on the parameters of the front of the drawer, as shown in Figure 15.6. This relationship will remain until you change it and update all the parts that share those parameters. This is one way of building parts to connect, but the changes are limited to manual updating. iParts will provide a more powerful technique.

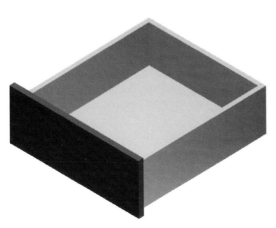

FIGURE 15.6 The drawer back repositioned by linking parameters

Building an iPart

You can think of an iPart as a part family—a part that includes a table so you can give different members of the part family different values to the same parameter or control whether a given feature appears in a family member.

In this exercise, you will build a second drawer front based on the one already in the file. You'll change its size and go from using a pull for opening to a knob.

1. Make certain that the 2013 Essentials project file is active, and then open c15-05.ipt from the Parts\Chapter 15 folder.

2. Change the Ribbon to the Manage tab, and click Create iPart in the Author panel.

   When the iPart Author dialog box opens (Figure 15.7), take a moment to click through some of the tabs that select which features will be controlled as part of the iPart. Notice that columns for the named parameter and a suppressed feature were automatically displayed and that -01 has been added to the current filename to define the first family member's part number.

3. Right-click the row of features for the current part, and click Insert Row in the context menu.

   This new member is given a -02 extension and currently has the same values as the original.

4. Change the Width value for the new member to 20 in and the Height value to 7 in.

> You can specify part numbers for each member, or you can click Options to open a dialog box where you can set the naming scheme.

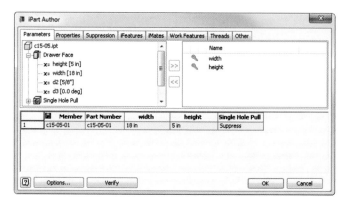

FIGURE 15.7 Beginning to create an iPart

5. Click the Suppression tab, find the Double Hole Pull [Compute] feature, and double-click it to add it to the option columns in both parts.

6. In the c15-05-02 row, change the value for the Double Hole Pull to 0.

7. In the same row, change the value for Single Hole Pull to Yes.

8. Compare your dialog box to Figure 15.8, and click OK to create the iPart.

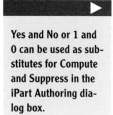

Yes and No or 1 and 0 can be used as substitutes for Compute and Suppress in the iPart Authoring dialog box.

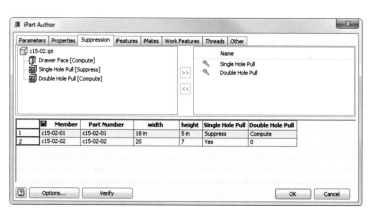

FIGURE 15.8 Adding a new size, swapping a feature

9. A new node has been added to the Browser. Expand the table to see that there are now callouts for the -01 and -02 versions of the part.

10. Double-click the c15-05-02 part, and select Zoom All to see the part.

11. Activate the c15-05-01 version.

12. In the Author panel of the Manage tab, find the Edit Factory scope icon, and use the drop-down to change the mode to Edit Member scope.

13. Add 0.75 in fillets to the short edges of the drawer front, as shown in Figure 15.9.

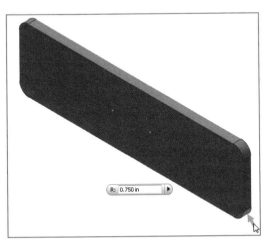

FIGURE 15.9 Adding a fillet to one of the iPart members

14. Switch the part to the c15-02-02 version to see that the fillets were not applied.

15. Double-click the table to see the effect of adding a feature with Member Scope active (Figure 15.10).

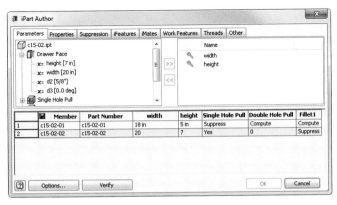

FIGURE 15.10 With Member Scope active, new features are not added to other members.

Perusing the tabs in the dialog box will give you a sense of just how many variables these models can have. Think about the numbers of variables in your designs that are similar. It's easy for the number of variables between parts to get out of hand.

Working with an Assembly of iParts

To configure an assembly, you can use a combination of regular parts or iParts. In each version of the assembly, you can change out what components are being used and how they are assembled to one another.

The assembly you will convert was built using traditional assembly constraints. The -01 version of each of the iParts was used in the assembly.

1. Make certain that the 2013 Essentials project file is active, and then open c15-02.iam from the Assemblies\Chapter 15 folder.

2. Locate the c15-11-01 part in the Browser, and expand it so you can see the Table icon.

3. Right-click the Table icon, and click Change Component in the context menu.

   The Place Standard iPart dialog box (Figure 15.11) appears, displaying the current version of the iPart you're using.

FIGURE 15.11 The Place Standard iPart dialog box displays the current iPart information.

4. In the dialog box, click the 2394-01 part number.

5. This will open a mini-dialog box showing the available versions of the part. Click 2394-03, and click OK.

   Because the holes for mounting the c15-16 part are suppressed in this version, there will be an error message concerning the failure of several assembly constraints.

6. Click Accept to continue working.

7. Locate the four failed constraints in the Browser under the c15-11-03 part.

8. Click one, and hold the Shift key to select the others. When they're all selected, right-click and click Suppress in the context menu.

9. Find the c15-17 part in the Browser and make it visible.

10. Create an insert constraint between the small hole in the drawer face and the bottom of the knob, as shown in Figure 15.12, clicking OK to place the constraint.

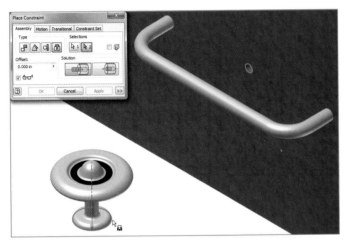

FIGURE 15.12 Locating the knob for the optional drawer front

11. Expand the c15-17 part, and rename the insert constraint Knob Insert constraint.

12. Orbit the assembly so you can see the back for the drawer front.

13. Add another insert constraint to the screw on the right and the hole in the drawer front, as shown in Figure 15.13.

14. In the Browser, expand the first instance of the Cross Recessed Truss Head screw.

15. Rename the Insert constraint at the bottom of the list to Knob Screw constraint.

16. Unsuppress the four constraints you suppressed under part c15-11-03, and accept their failure in the warning dialog box.

17. Use the Change Component tool to change the c15-11-03 part back to c15-11-01.

18. A warning dialog box will appear, letting you know that two constraints have failed. Accept this.

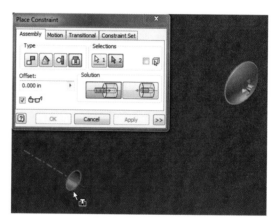

FIGURE 15.13 Reusing a screw by applying another constraint

Now the assembly is prepared for a conversion to an iAssembly. It is not absolutely necessary to create the constraints for the knob ahead of time. It is helpful to build several configurations in one step, but for your products, you might choose to create iAssemblies one member at a time, adding parts as you go.

Converting an Assembly to an iAssembly

All the pieces are in place, and the assembly constraints are ready to be used. Naming the constraints has the same benefit as naming parameters in an iPart, but it makes them much easier to locate in the iAssembly Author dialog box.

1. Make certain that the 2013 Essentials project file is active, and then open c15-03.iam from the Assemblies\Chapter 15 folder.

Create iAssembly

2. Click Create iAssembly in the Author panel of the Manage tab.
The iAssembly Author dialog box (Figure 15.14) opens, and as you might expect, it will look a lot like the iPart Author dialog box. The main focus of this dialog box is to specify which iPart member to show in the assembly and whether any constraints need to change.

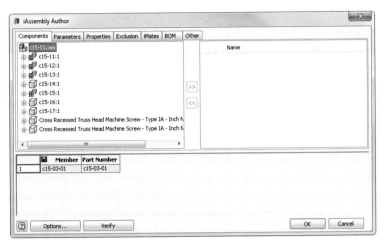

FIGURE 15.14 The initial appearance of the iAssembly Author dialog box

3. On the Components tab, expand the c15-11 part.

4. Double-click Table Replace to add it to the pane on the right, and add it as a column for selection for the -01 version of the assembly.

5. Repeat the previous step for the c15-12, c15-13 (both instances), and c15-15 parts. See Figure 15.15.

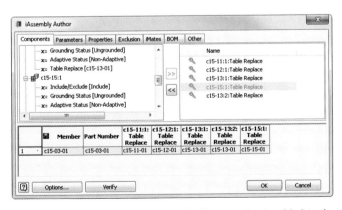

FIGURE 15.15 Control over the iPart version is added to the iAssembly member.

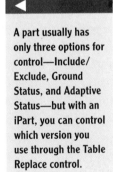

A part usually has only three options for control—Include/ Exclude, Ground Status, and Adaptive Status—but with an iPart, you can control which version you use through the Table Replace control.

6. Switch to the Exclusion tab, and expand the Components folder.

7. Double-click c15-16 and c15-17 to control whether they will be in the assembly.

8. In the column for c15-17, click the Include value, and use the drop-down to change it to Exclude.

9. Add the last instance of the Cross Recessed Truss Head screw as well.

10. Now expand the Constraints folder on the Exclusion tab, and add the six named assembly constraints to the list.

11. Set the values for the two constraints for the knob to Exclude.

12. Right-click the iAssembly row, click Insert Row to add another member, and then add a third using Insert Row.

13. Change the iPart version of each iPart to match the member version of the assembly.

14. For iAssembly Member -03, exclude the c15-16 part and the Cross Head screw.

15. Then exclude the constraints that have *Wire Pull* in the name.

16. Set the c15-17 part and constraints that say *Knob to Include*.

17. Set the Part Number names to 24 inch for -01, 20 inch for -02, and 16 inch for -03.

    Since this iAssembly might be used in other assemblies, it's important to make it easy to click the version you want to use. Applying a key to the features or information that an end user might need makes it much easier for others to find the right iAssembly member.

18. Right-click in the Part number column, and hover over Key in the context menu until a list of numbers appears. Click 1 in the list to make the part number the first (and only, in this case) key for selecting a member of the iAssembly.

19. Compare all the settings in your dialog box to Figure 15.16.

20. Click OK to generate the iAssembly.

21. In the assembly, expand the Table folder near the top of the Browser.

22. Double-click the instances to see the versions of the assembly.

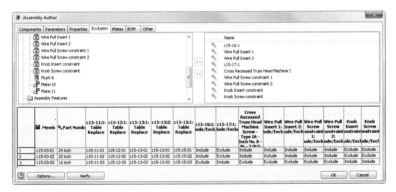

FIGURE 15.16 An example of how many options can be controlled in
an iPart or iAssembly

Compared with your products, this was probably a simple assembly. Managing
the contents of an iAssembly can seem a bit daunting, but having the ability to
access variations on a common assembly through a single assembly file is great
for maintaining the design.

Using an iAssembly

The primary benefit of having an iAssembly is to have the option of placing different versions of an assembly family into the same primary assembly. The version inserted into the assembly can also be changed at any time.

1. Make certain that the 2013 Essentials project file is active, and then
    create a new assembly using the Standard.(mm).iam template from
    the Metric folder.

2. Start the Place tool in the marking menu or the Component panel on
    the Assembly tab.

3. In the Open dialog box, double-click the c15-04.iam from the
    Assemblies\Chapter 15 folder.
    Selecting an iAssembly will preview the component and display
    the Place iAssembly dialog box. You can use this dialog box to select
    which member of an iAssembly you want to place. The same type of
    dialog box appears when you place an iPart into an assembly.

4. Set the Graphics window to the Home view and click the screen to
    place the default version of the drawer.

5. After the screen shows the first instance, place two more drawers in the assembly, and then click Dismiss or hit the Esc key to close the dialog box. See Figure 15.17.

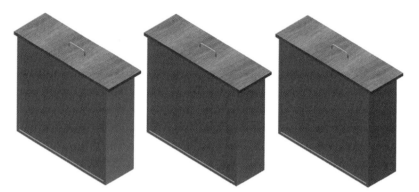

FIGURE 15.17 Three instances of the same iAssembly placed in an assembly

6. In the Browser, expand the second instance of the assembly.

7. Right-click the Table icon, and click Change Component in the context menu.

8. Click the current 24 inches value, and select 20 inch from the mini-dialog box; then click OK.

The assembly will update to the new version. In some cases, it might prompt you to save the new file; if so, just click OK.

9. Now use the Change Component tool for the third instance.

10. When the dialog box opens, click the Tree tab to see a list of the size options.

11. Click the Table tab to see a view of the entire configuration table.

12. Click the bottom row; then click OK to update the assembly to that version.

When your screen updates to show the new version of the drawer (Figure 15.18), you can see the differences between the configurations side by side. This is a common application for iAssemblies, where subassemblies are far more likely to have modules with options. Another feature is that in many cases, changing the family member doesn't require the reapplication of assembly constraints.

FIGURE 15.18 The three iAssemblies showing different states

Documenting an iPart and an iAssembly

Combining iParts and iAssemblies with the Inventor drawing associativity feature is a great way to simplify maintaining these documents. In this exercise, you will edit a drawing to include a table for an iPart and a specialized parts list that shows the mixture of parts used in an iAssembly.

1. Make certain that the 2013 Essentials project file is active, and then open c15-01.idw from the Drawings\Chapter 15 folder.

2. Click the Table tool in the Table panel of the Annotate tab.

3. Click the drawing view of the drawer front.
   Once a view of an iPart is selected, the Table dialog box adjusts to show a column for the members.

4. Click the Column Chooser tool in the dialog box to open another dialog box, where you can select the properties of the iPart to include in the table.

5. In the Table Column Chooser dialog box, add drawer_width, drawer_height, and dbl_pull_width to the Selected Columns group on the right (Figure 15.19); then click OK.

6. Click OK again to begin placing the table in the drawing.

7. Place the table under the drawing view.

8. Double-click the table to open the editing dialog box, right-click the drawer_width column, and click Format Column from the context menu.

9. In the Format Column dialog box, uncheck Name From Data Source, and then change the Heading name to A, as shown in Figure 15.20.

10. Repeat the process to change the heading of the drawer_height column to B and dbl_pull_width to C.

11. Click OK to close the dialog box and update the table, as shown in Figure 15.21.

    With the iPart table update, you can add a parts list for the iAssembly. The parts list for the assembly needs to be able to display quantities for any component that is a member of any of the various configurations. This is sometimes referred to as a *matrix parts list*. The Parts List tool has a special function to do this.

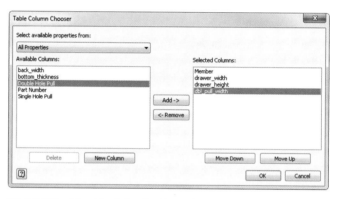

FIGURE 15.19 Selecting items for the iPart table

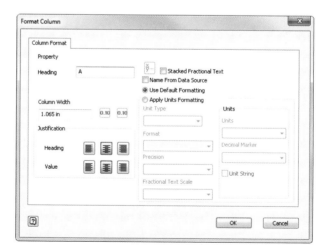

FIGURE 15.20 Modifying the Column header

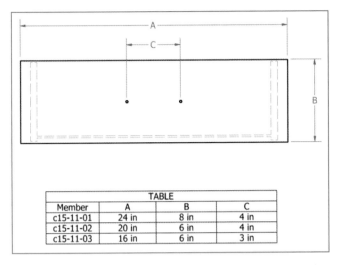

TABLE			
Member	A	B	C
c15-11-01	24 in	8 in	4 in
c15-11-02	20 in	6 in	4 in
c15-11-03	16 in	6 in	3 in

FIGURE 15.21 The completed table showing the values for the table dimensions

12. Find the Parts List tool on the Annotate tab in the Table panel, and click it.

13. Click the drawing view of the assembly, and then click OK to place a standard parts list under the assembly view.

14. Double-click the parts list to open the editing dialog box.

15. Click the Member Selection icon.

16. In the Select Member dialog box, click all three versions of the iAssembly, and click OK.

17. Click the Column Chooser button, and remove the Description column from the Selected Properties list.

18. Click OK twice to update the parts list on the drawing.

As Figure 15.22 shows, the parts list contains the quantity of the parts in rows and the configuration of the assembly in columns.

For many years, "family parts" or configurations of an assembly could be included on 2D drawings only as tables. These tables could be very complex and include a lot of information, and if a change to a family member occurred, it was easy to miss it in the drawing.

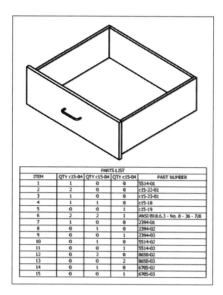

| | | PARTS LIST | | |
ITEM	QTY c15-04	QTY c15-04	QTY c15-04	PART NUMBER
1	1	0	0	5514-01
2	2	0	0	c15-22-01
3	1	0	0	c15-23-01
4	1	1	0	c15-18
5	0	0	1	c15-19
6	2	2	1	ANSI B18.6.3 - No. 8 - 36 - 7/8
7	1	0	0	2394-01
8	0	1	0	2394-02
9	0	0	1	2394-03
10	0	1	0	5514-02
11	0	0	1	5514-03
12	0	2	0	8650-02
13	0	0	2	8650-03
14	0	1	0	6705-02
15	0	0	1	6705-03

FIGURE 15.22 The parts list shows the quantities of parts for each iAssembly member.

Expanding the Control Options

Using iParts and iAssemblies gives a good level of control over well-defined components, but it is limited to members that were predefined. If you want something unique, you have to build a new member.

The iLogic tools work with many of the same parameters but offer a way to programmatically change these parameters based on the condition of other parameters. Using iLogic tools, you can also establish a range of sizes rather than strictly working with a predefined set.

In the next exercises, you will do the same type of functions as you did with iAssemblies and iParts, but you will use the iLogic interface to establish connections in a more meaningful way and create a dialog box that will make it easy for you to edit the product.

Using a Parameter in Another Parameter

To make sure the solid part of the drawer is cut to the proper size, you need to allow for the thickness of the edge banding material. To do this, you can build an equation into the value of the parameter.

> **Many of these exercises transition from one to the next, so you will be asked to save your work and continue using the same data for several exercises.**

1. Make certain that the 2013 Essentials project file is active, and then open c15-18.ipt from the Parts\Chapter 15 folder.

2. In the Parameters panel of the Manage tab, click the Parameters tool.

3. In the lower-left corner of the Parameters dialog box, click the filter, and set the dialog box to show only renamed parameters.

   At the top of the Model Parameters list is a parameter named drawer_core_width, whose current value is 20 in. This is the parameter whose value you will modify.

4. Click the cell in the Equation column, and type **drawer_width-(edge_band_thickness * 2 ul)**; press Enter to update the dialog box. See Figure 15.23.

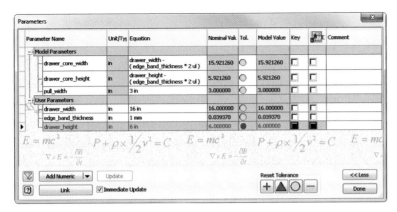

FIGURE 15.23 Parameters can be factored together to create a value for another parameter.

If you can see the part in the background, it will immediately change size. You can disable this for large models with the check box at the bottom of the dialog box. The *ul* in the equation tells Inventor that the number is strictly a number and not a dimensional value.

5. Click Done to close the dialog box.

6. Be sure to save your changes and close the part.

Now, your part will maintain the proper size when changes to the overall drawer width or thickness of the edge banding occur.

Creating a Multivalue Parameter

It's possible to set up any parameter local to a part to use a list of predefined values rather than typing in a value, but it's when you start working with the iLogic tools that the capability really shows its benefit.

1. Make certain that the 2013 Essentials project file is active, and then open c15-05.iam from the Assemblies\Chapter 15 folder.

2. In the Manage panel of the Assemble tab, start the Parameters tool.

3. Click the Add Numeric icon to add a new parameter to the model.

4. The new row will appear with the cursor ready to give it a name. Type edge_band_thickness, and press Enter.

5. In the Unit column, click the field next to the new parameter.

6. You can browse under the Length category in the Unit Type dialog box, or you can just type **mm** for the Unit Specification value and click OK.

7. Right-click the new parameter, and click Make Multi-Value in the context menu.

The Value List Editor will appear, showing the current value (1.0 in) listed in the bottom portion of the dialog box.

▶

In addition to setting a list of sizes, you can allow custom values to be used by selecting the check box in the Value List Editor dialog box.

8. Click this value, and then click the Delete button on the right to clear the list.

9. In the Add New Items field at the top of the dialog box, type 1, press the Enter key, and then add 3 to the list.

10. Click the Add button to put these values on the list shown in Figure 15.24.

FIGURE 15.24 Creating a defined list of possible parameter values

11. Click OK to close the dialog box and go back to the Parameters dialog box.

12. Click the drop-down in the Equation column, and set the value for the parameter to 1 mm; then click Done to close the dialog box.

13. Save the assembly, but do not close it.

You've now established a parameter in the assembly that will control the dimensions on two parts and modify the way a third part calculates a dimension. Furthermore, you've limited the sizes to standard materials.

Accessing the iLogic Tools

This assembly already has a couple of rules included to give you a head start. To see them, you need to turn on the iLogic Browser. If you did not do the previous exercise, the following steps will fail.

Certification Objective

1. Continue using the c15-05.iam file from the previous exercise.

2. Switch to the Manage tab, and click the iLogic Browser tool in the iLogic panel.

   iLogic Browser

   The iLogic Browser (Figure 15.25) appears and might be attached to your model Browser. On the Rules tab, you will see the two rules that were created in advance.

FIGURE 15.25 Clicking and dragging the header can make the iLogic Browser a floating window.

3. Right-click Edge Band Control, and select Unsuppress Rule from the context menu.

4. Right-click the rule again, and click Run Rule in the context menu.

5. The update icon will appear in the Quick Access toolbar as a result of parameters being updated. Look closely at the thickness of the edge band, and click Update.

6. Save the assembly.

The four bands will take on the value of the edge_band_thickness parameter. In a later exercise, you will change the thickness. Now, you need to build rules to make sure the edge banding is the proper length.

Creating a New Rule

Rules are simply ways to establish behavior or to create conditions in which an action will be taken if another action creates a condition. The rule you will create could be built as several rules; or, with good use of commenting, it can be handled in one rule.

1. Make certain that the 2013 Essentials project file is active, and then open c15-06.iam from the Assemblies\Chapter 15 folder.

2. Right-click in the iLogic Browser, and click Add Rule in the context menu.

3. When the Rule Name dialog box appears, type **Drawer Size Control**, and click OK.

   A large Edit Rule dialog box will open. The first task in this new rule will be to limit the size of a drawer front.

4. With the Model tab of the dialog box active, click the User Parameters icon under the assembly to display the user parameters in the assembly.

   Setting a maximum and minimum size range is made easy thanks to a wizard included in iLogic.

Parameter Limits

5. Click the Parameter Limits icon on the Wizard tab of the Edit Rule dialog box.

6. In the iLogic Limits Wizard dialog box, enter the parameter name of **drawer_width**, and set the Max. Value setting to 34 and the Min. Value setting to 10, as shown in Figure 15.26.

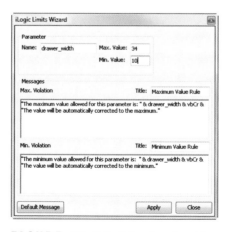

FIGURE 15.26 Using the wizard to set a size range for the drawer front

You can add custom messages for violation of the minimum or maximum values in the dialog box as well.

7. Click Apply to create the width control portion of the rule.

8. Change the parameter name to drawer_height, and set the Max. Height value to **12** and the Min. Height value to **4**.

9. Click Apply and Close to see the rule up to this point in Figure 15.27.

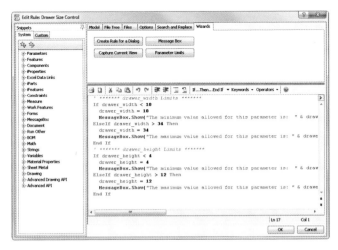

FIGURE 15.27 Building rules to limit the size of a drawer face

Let's separate this part of the rule from the next part with a comment. To create a comment as a note or to remove a line temporarily from the rule, place an apostrophe at the beginning of the row.

10. Below the last line containing End If, press the Enter key a couple of times, and then type **This section is for linking assembly parameters to part parameters.**

This will make for a clean break and help you sort out what portion of the rule is controlling what functions or relationships in the rule.

11. Click OK. Save the assembly.

Now, you need to make the connections to the components of the assembly so that the size limitations can affect them.

Linking the Parameters Through Rules

In the first portion of this chapter, you learned how to link parameters between models by using a spreadsheet or by linking directly to another component. The next step does the same thing, but by using the lines of code, it can be set up very quickly.

1. Continue using the c15-06.iam file from the previous exercise.

2. In the iLogic Browser, right-click Drawer Size Control rule, and select Edit Rule.

3. Below the comment added in the previous section, type the following: Parameter ("c15-21:1," "drawer_width") = drawer_width.

   This statement simply says that the parameter named drawer_width in the first assembly instance of the model named c15-21 gets its value from the parameter named drawer_width in this assembly.

4. Highlight that line, copy it to the clipboard, and paste a copy directly below this one. It's a good idea to create a little space between lines.

5. In the new line, change the word **width** to **height** so that it reads as follows: Parameter ("c15-21:1," "drawer_height") = drawer_height.

   Now, you've linked the height value of the assembly to the part. Next, you need to leverage these values with the edge banding. Even if you don't set up every parameter name in one component to match up to the parameter names of another, you can still create effective relationships between them.

6. Paste the text into the window again and modify it to say the following: Parameter ("c15-22:1," "length") = drawer_width.

   The last stage in the rule has an extra step because you need to subtract the thickness of the edge band, as you did when creating the equation in the Parameters table.

7. Create a new row of text to read as follows: Parameter ("c15-23:1," "length") = (drawer_height - (edge_band_thickness) *2).

8. Compare your rules to Figure 15.28, and click OK to save the rule you've created.

```
' This section is for linking assembly parameters to part parameters

Parameter ("c15-21:1", "drawer_width") = drawer_width

Parameter ("c15-21:1", "drawer_height") = drawer_height

Parameter ("c15-22:1", "length") = drawer_width

Parameter ("c15-23:1", "length") = (drawer_height - (edge_band_thickness) *2)
```

FIGURE 15.28 Rules can create connections between components.

9. To update the assembly, click the Update icon on the Quick Access toolbar.

10. After the assembly updates, save the file.

When the assembly updates, you will see that the edge band components are now the correct size. Once you become comfortable with the language iLogic speaks, you will see that you're just giving orders based on needs that already exist around your designs.

Controlling Features Through Rules

Now, you have your file parameters connected to each other, and the assembly is looking good. There is one last step before you test all the connections. There needs to be a rule that changes the number of holes when the drawer becomes narrow. Rather than continuing to test your typing skills, this exercise will leverage the ability to store rule content outside of the Inventor file:

1. Continue using the c15-06.iam file from the previous exercise.

2. Use Windows Explorer to open iLogic Feature Suppression.txt from the Assemblies\Chapter 15 folder.

3. When the text file opens, press Ctrl+A to select the entire contents, and then copy the selected text to the Clipboard by pressing Ctrl+C.

4. Switch back to Inventor.

5. Double-click the Pull Control rule in the iLogic Browser to open it in the editor.

6. Press the Enter key to create space beneath the existing contents in the rule, and paste the information you copied from the text file, as shown in Figure 15.29.

You'll notice that the actions and parameter names are automatically color-coded even if you didn't type the rule in the window. The rule says that if the value of drawer_width is equal to or less than 14 inches, then Inventor will suppress the Double Hole Pull feature and unsuppress the Single Hole Pull feature. If the drawer is larger than 14 inches, it will place two holes in the part.

7. Click OK to update the rule and close the dialog box.

The Edge Band Control rule has a feature suppression function built in that suppresses a fillet on the edge band when the edge is 1 mm thick and turns it on when it is 3 mm.

```
' ** This section switches the panel to one hole if 14 inches wide or less

If drawer_width < 14 in

' ***Single Hole Pull***
Feature.IsActive("c15-21:1", "Single Hole Pull") = True

' ***Double Hole Pull***
Feature.IsActive("c15-21:1", "Double Hole Pull") = False

ElseIf drawer_width = 21 in

' ***Single Hole Pull***
Feature.IsActive("c15-21:1", "Single Hole Pull") = True

' ***Double Hole Pull***
Feature.IsActive("c15-21:1", "Double Hole Pull") = False

ElseIf drawer_width > 14 in

' ***Single Hole Pull***
Feature.IsActive("c15-21:1", "Single Hole Pull") = False

' ***Double Hole Pull***
Feature.IsActive("c15-21:1", "Double Hole Pull") = True

End If
```

FIGURE 15.29 Features can be switched on and off by rules.

As you continue to expand the application of iLogic for your designs, you'll also see that there is a tremendous amount of sample code available online and in the product. All you need to do is swap out file, feature, or parameter names as you've already done, and you're up and running.

Making the Rules Easy to Use

Any of the rules you create can be tested and used by changing a couple of parameters in the dialog box, but that is still cumbersome for many users. Using a basic form to open and use with the tool is a great step forward in usability.

1. Make certain that the 2013 Essentials project file is active, and then open c15-07.iam from the Assemblies\Chapter 15 folder.

2. Open the iLogic Browser, and click the Forms tab.

3. Click the Drawer Front Size button.
   The Drawer Front Size dialog box opens in the Design window. You can test it on your assembly.

4. Set Drawer Width to 12 in and Drawer Height to 3 in, and click Apply.
   An error dialog box appears (Figure 15.30), warning that the height was below the minimum you specified in the parameter limits function of the Drawer Size Control rule.

5. Click OK so the assembly can adjust to the width entered and the minimum height.
   Note that there is only one hole in the part because it is less than 14 inches wide.

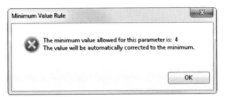

FIGURE 15.30 An error dialog box was built directly from the wizard used to set size limits.

6. Change the width to 16 and the height to 7, click the Edge Band tab, click the radio button for 3 mm Edge Band Thickness, and click Apply.

   The fillet develops on the thicker edge band. Now, you need to add to this form the option of using two different size pull spacing.

7. Right-click the button on the Forms tab, and click Edit in the context menu.

   In the Form Editor dialog box, you can browse the parameters available in the file and elements for making a dialog box. First you should add a tab to place controls for the pull size.

8. Click and drag the Tab group from the Toolbox group onto the Drawer Front Size text in the top-right window.

9. Change the name of the new Tab group to Pull Width.

10. Drag a label under the Pull Width tab, and set it to read Select Pull Width.

11. Locate the pull_width parameter on the Parameters tab, and drag it under the label.

12. Click the new item, and the window below it will change to display the properties.

13. Click Edit Control Type, and change it from Combo Box to Radio Group using the drop-down, as shown in Figure 15.31.

14. If you don't see a preview on your screen, click the Preview button in the dialog box, and click the new Pull Width tab.

15. Click OK to update the form.

16. Click the Drawer Front Size button again to test the new tab.

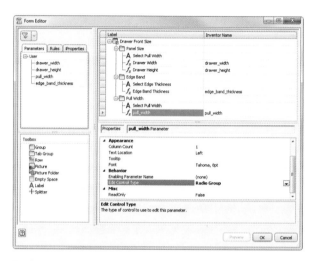

FIGURE 15.31 Editing the form to add control over the size of the pulls

You can experiment with different ways of using the form too. One great example is to replace the Text Box input for the drawer width and height with a slider. When you set the maximum and minimum values and the size of the increment, another layer of control over using the parameter value range in the rules is added.

The tools I've covered are just the essentials of iLogic, but learning how to do this much should encourage you to explore how it might help you by removing tedium and repetition from your daily work. There are some amazing examples of what users with little programming experience have done using iLogic and Inventor.

THE ESSENTIALS AND BEYOND

Automation of design tasks is why many people moved to CAD and 3D CAD tools such as Inventor in the first place. That's why it's amazing how many users are just beginning to notice and take advantage of tools such as iParts, iAssemblies, and iLogic. Look around for something simple that has a lot of size variation but relatively few changes in the number or type of parts, and it will make a great project for exploring the tools of this chapter.

THE ESSENTIALS AND BEYOND *(Continued)*

Any time you can reduce the number of files you need to maintain, you reduce the cost of searching for data and the risk of parts with mismatched feature sizes. Using tables to control and understand your data is an elevated approach to parametric modeling and assembly control. Further still in the evolution of computer-aided design is implementing the rules that safeguard the quality of your product to produce them.

ADDITIONAL EXERCISES

▶ Edit a part using the values in the tables rather than editing the sketch or features.

▶ Construct an iAssembly of parts that are similar but are not iParts, and use the table to swap them for each other.

▶ Try using iLogic to build rules to change materials more rapidly than the Style and Standards Editor.

▶ Download Microsoft Visual Basic Express (which is free), and build executable files out of your iLogic rules.

Autodesk Inventor Certification

Autodesk certifications are industry-recognized credentials that can help you succeed in your design career, providing benefits to both you and your employer. Getting certified is a reliable validation of skills and knowledge, and it can lead to accelerated professional development, improved productivity, and enhanced credibility.

This Autodesk Official Training Guide can be an effective component of your exam preparation. Autodesk highly recommends (and we agree!) that you schedule regular time to prepare by reviewing the most current exam preparation road map available at www.autodesk.com/certification, using Autodesk Official Training Guides like this one, taking a class at an Authorized Training Center (find them near you here: www.autodesk.com/atc), and using a variety of other resources—including plenty of hands-on experience.

To help you focus your studies on the skills you'll need for these exams, the following tables show objectives that could potentially appear on an exam and in which chapter you can find information on that topic—and when you go to that chapter, you'll find certification icons like the one in the margin here.

Certification Objective

Table A.1 provides information for the Autodesk Certified Associate exam and lists the section, exam objectives, and chapter where the information is found. Table A.2 provides information for the Autodesk Certified Professional exam. The sections and exam objectives listed in the table are from the Autodesk Certification Exam Guide.

These Autodesk exam objectives were accurate at press time; please refer to www.autodesk.com/certification for the most current exam road map and objectives.

Good luck preparing for your certification!

TABLE A.1 Autodesk Certified Associate exam sections and objectives

Topic	Learning Objective	Chapter
User Interface: Primary Environments	Name the four primary environments: Parts, Assemblies, Presentations, and Drawings.	Chapter 1

(Continues)

TABLE A.1 *(Continued)*

Topic	Learning Objective	Chapter
User Interface: UI Navigation/ Interaction	Name the key features of the user interface (ribbon ➢ panels ➢ tabs).	Chapter 1
	Describe the listing in the browser for an assembly file.	Chapter 1
	Demonstrate how to use the context (right-click) menus.	Chapter 1
	Demonstrate how to use the menus.	Chapter 1
	Demonstrate how to add Redo to the Quick Access Toolbar.	Chapter 1
User Interface: Graphics Window Display	Describe the steps required to change the background color of the graphics window.	Chapter 1
	Use Application Options ➢ Display.	Chapter 1
	Demonstrate how to turn on/off the 3D Indicator.	Chapter 1
User Interface: Navigation Control	Describe the functionality of the ViewCube.	Chapter 1
	Describe the Navigation bar.	Chapter 1
User Interface: Navigation Control	Name the navigation tools started by the F2 to F6 shortcut keys: Pan (F2), Zoom (F3), Free Orbit (F4), Previous View (F5), Home View (F6).	Chapter 1
File Management: Project Files	Name the file extension of a project file (.ipj).	Chapter 1
	List the types of project files that can be created.	Chapter 1
	Define the term workspace.	Chapter 1
	List the types of files stored in a library.	Chapter 1
	List the three categories in Folder Options.	Chapter 1
	Describe how to set the active project.	Chapter 1
Sketches: Creating 2D Sketches	Name the file extension of a part file (.ipt).	Chapter 1
	Describe the purpose of a template file in the sketch environment.	Chapter 5

TABLE A.1 *(Continued)*

Topic	Learning Objective	Chapter
	Describe the function of the 3D Coordinate System icon.	Chapter 1
	Define a sketch plane.	Chapter 3
	Label the entries on the browser.	Chapter 1
Sketches: Draw Tools	Complete a 2D sketch using the appropriate draw tools: Line, Arc, Circle, Rectangle, Point, Fillet, Polygon.	Chapters 3, 7
Sketches: Sketch Constraints	List the available geometric constraints: coincident, colinear, concentric, fixed, parallel, perpendicular, horizontal, vertical, tangent, symmetric, and equal.	Chapters 3, 7
	Describe parametric dimensions (general and automatic).	Chapters 3, 7
	Describe how to control the visibility of constraints.	Chapter 3
	Describe the degrees of freedom on a sketch and how they can be displayed.	Chapter 3
Sketches: Pattern Sketches	Demonstrate how to pattern a sketch (rectangular, circular, and rotate).	Chapter 3
Sketches: Modify Sketches	Demonstrate how to move a sketch.	Chapter 3
	Demonstrate how to copy a sketch.	Chapter 3
	Demonstrate how to rotate a sketch.	Chapter 3
	Demonstrate how to trim a sketch.	Chapter 3
	Demonstrate how to extend a sketch.	Chapter 3
	Demonstrate how to offset a sketch.	Chapter 3
Sketches: Format Sketches	Describe how to format sketch linetypes.	Chapter 3
	Discuss over constrained sketches.	Chapter 3
Sketches: Sketch Doctor	Examine a sketch for errors.	Chapter 3

(Continues)

TABLE A.1 *(Continued)*

Topic	Learning Objective	Chapter
Sketches: Shared Sketches	Describe the function of a shared sketch.	Chapters 3, 7, 9, 10
Sketches: Sketch Parameters	Describe how parameters define the size and shape of features.	Chapter 3
Parts: Creating Parts	Name the file extension of a part file (.ipt).	Chapter 1
	Label the entries on the browser.	Chapter 1
	Define a base feature.	Chapter 3
	Define an unconsumed sketch.	Chapter 3
	Demonstrate how to create an extruded part.	Chapter 3
	Demonstrate how to create a revolved part.	Chapter 3
	Demonstrate how to create a lofted part.	Chapter 7
	Describe the termination options for a feature.	Chapters 3, 7
	Demonstrate how to create a hole feature.	Chapters 3, 7
	Demonstrate how to create a fillet feature.	Chapters 3, 7
	Demonstrate how to create a chamfer feature.	Chapter 3
	Demonstrate how to create a shell feature.	Chapter 7
	Demonstrate how to create a thread feature.	Chapter 7
Parts: Work Features	Describe the use of work features (plane, point, axis) in the part creation workflow.	Chapter 7
Parts: Pattern Features	Demonstrate how to create a rectangular pattern.	Chapter 3
	Demonstrate how to create a circular pattern.	Chapter 3
	Demonstrate how to mirror features.	Chapter 7

TABLE A.1 *(Continued)*

Topic	Learning Objective	Chapter
Parts: Part Properties	Describe part properties and how they are applied.	Chapters 3, 5
Assemblies: Creating Assemblies	Name the file extension of an assembly file (.iam).	Chapters 1, 4, 8, 11, 12, 14, 15
	Label the entries on the browser.	Chapter 1
	Name the six degrees of freedom on a component.	Chapter 4
	Demonstrate how to place a part in an assembly.	Chapter 4
	Discuss degrees of freedom and a grounded part.	Chapter 4
	Demonstrate how to apply various assembly constraints.	Chapters 4, 8
	Describe the various assembly environment techniques (top-down, bottom-up, and middle-out).	Chapters 4, 8
	Demonstrate how to create a new part in the assembly environment.	Chapter 8
	Demonstrate how to place a Content Center part in an assembly.	Chapters 4, 8
Assemblies: Animation Assemblies	Demonstrate how to animate an assembly using drive constraints.	Chapter 8
Assemblies: Adaptive Features, Parts, and Subassemblies	Demonstrate how to make and use an adaptive part.	Chapter 8
Presentations	Name the file extension of a presentation file (.ipn).	Chapters 1, 13, 14
	Label the entries on the browser.	Chapter 1

(Continues)

TABLE A.1 *(Continued)*

Topic	Learning Objective	Chapter
	Discuss the various uses of Presentation files.	Chapters 6, 13
	Demonstrate how to apply tweaks to a part.	Chapter 13
	Demonstrate how to apply trails to a part.	Chapter 13
	Demonstrate how to animate an assembly.	Chapter 13
Drawings	Name the file extension of a drawing file (.idw).	Chapters 1, 2, 5, 6, 10, 14, 15
	Describe the use of template files.	Chapter 1, 5
	Label the entries on the browser.	Chapter 1
	Describe the content within Drawing Resources.	Chapter 5
	Demonstrate how to create a part drawing.	Chapters 2, 6, 10
	Demonstrate how to create an assembly drawing.	Chapter 6
	Describe the various annotation options.	Chapters 2, 6, 10
	Demonstrate how to add balloons to an assembly.	Chapter 6
Sheet Metal: Creating Sheet Metal Parts	Name the file extension of a sheet metal part file (.ipt).	Chapter 10
	Discuss the use of sheet metal defaults.	Chapter 10
	Demonstrate the creation of a sheet metal bend. Use Create Tools ➢ Bend, Face, and Flange.	Chapter 10
Sheet Metal: Modify Sheet Metal Parts	Demonstrate the creation of a corner seam.	Chapter 10
	Demonstrate the creation of a punch tool.	Chapter 10
	Demonstrate the creation of a cut across a bend.	Chapter 10

TABLE A.1 *(Continued)*

Topic	Learning Objective	Chapter
Sheet Metal: Flat Pattern	Demonstrate how to create a flat pattern.	Chapter 10
	Demonstrate how to insert a flat pattern in a drawing.	Chapter 10
	Demonstrate how to export a flat pattern.	Chapter 10
Visualization: Create Rendered Images	Describe the process to activate Inventor Studio.	Chapter 13
	Demonstrate how to create a new camera.	Chapter 13
	Demonstrate how to create a rendered image.	Chapter 13
Visualization: Animate an Assembly	Demonstrate how to create a new animation.	Chapter 13
	Demonstrate how to create an animation by animating a camera.	Chapter 13
	Demonstrate how to create an animation by animating a constraint.	Chapter 13
	Demonstrate how to create an animation by animating a fade.	Chapter 13

TABLE A.2 Autodesk Certified Professional exam sections and objectives

Topic	Learning Objective	Chapter
Advanced Modeling	Create a 3D path using the Intersection Curve and the Project to Surface commands.	Chapters 2, 3, 7, 15
	Create a loft feature.	
	Create a multi-body part.	
	Create a part using surfaces.	
	Create a sweep feature.	
	Create an iPart.	
	Create and constrain sketch blocks.	
	Use iLogic.	
	Emboss text and a profile.	

(Continues)

TABLE A.2 *(Continued)*

Topic	Learning Objective	Chapter
Assembly Modeling	Apply and use assembly constraints. Create a level of detail. Create a part in the context of an assembly. Describe and use Shrinkwrap. Create a positional representation. Create components using the Design Accelerator commands. Modify a bill of materials. Find minimum distance between parts and components. Use the frame generator command.	Chapters 4, 8, 11, 14
Drawing	Create and edit dimensions in a drawing. Edit a section view. Modify a style in a drawing. Edit a hole table. Modify a parts list. Edit a base and projected views.	Chapters 2, 5, 6
Part Modeling	Create a pattern of features. Create a shell feature. Create extrude features. Create fillet features. Create hole features. Create revolve features. Create work features. Use the Project Geometry and Project Cut Edges commands.	Chapters 3, 7
Presentation Files	Animate a presentation file.	Chapter 13
Project Files	Control a project file.	Chapter 1

TABLE A.2 *(Continued)*

Topic	Learning Objective	Chapter
Sheet Metal	Create flanges.	Chapter 10
	Annotate a sheet metal part in a drawing.	
	Create and edit a sheet metal flat pattern.	
	Describe sheet metal features.	
Sketching	Create dynamic input dimensions.	Chapters 3, 7, 10
	Use sketch constraints.	
User Interface	Identify how to use visual styles to control the appearance of a model.	Chapter 1
Weldments	Create a weldment.	Chapter 12

INDEX

Note to the Reader: Throughout this index **boldfaced** page numbers indicate primary discussions of a topic. *Italicized* page numbers indicate illustrations.

D